Textile Progress

September 2010
Vol 42   No 3

Geotextiles: production, properties and performance

**Amit Rawal, Tahir Shah
and Subhash Anand**

T0138866

The Textile Institute

Taylor & Francis
Taylor & Francis

SUBSCRIPTION INFORMATION

Textile Progress (USPS Permit Number pending), Print ISSN 0040-5167, Online ISSN 1754-2278, Volume 42, 2010.

Textile Progress (www.tandf.co.uk/journals/TTPR) is a peer-reviewed journal published quarterly in March, June, September and December by Taylor & Francis, 4 Park Square, Milton Park, Abingdon, Oxon, OX14 4RN, UK on behalf of The Textile Institute.

Institutional Subscription Rate (print and online): $335/£176/€267
Institutional Subscription Rate (online-only): $319/£168/€254 (plus tax where applicable)
Personal Subscription Rate (print only): $123/£63/€98

Taylor & Francis has a flexible approach to subscriptions enabling us to match individual libraries' requirements. This journal is available via a traditional institutional subscription (either print with free online access, or online-only at a discount) or as part of the Engineering, Computing and Technology subject package or S&T full text package. For more information on our sales packages please visit www.tandf.co.uk/journals/pdf/salesmodelp.pdf.

All current institutional subscriptions include online access for any number of concurrent users across a local area network to the currently available backfile and articles posted online ahead of publication.

Subscriptions purchased at the personal rate are strictly for personal, non-commercial use only. The reselling of personal subscriptions is prohibited. Personal subscriptions must be purchased with a personal cheque or credit card. Proof of personal status may be requested.

Ordering Information: Please contact your local Customer Service Department to take out a subscription to the Journal: **India**: Universal Subscription Agency Pvt. Ltd, 101–102 Community Centre, Malviya Nagar Extn, Post Bag No. 8, Saket, New Delhi 110017. **USA, Canada and Mexico**: Taylor & Francis, 325 Chestnut Street, 8th Floor, Philadelphia, PA 19106, USA. Tel: +1 800 354 1420 or +1 215 625 8900; fax: +1 215 625 8914; email: customerservice@taylorandfrancis.com. **UK and all other territories**: T&F Customer Services, Informa Plc., Sheepen Place, Colchester, Essex, CO3 3LP, UK. Tel: +44 (0)20 7017 5544; fax: +44 (0)20 7017 5198; email: subscriptions@tandf.co.uk.

Dollar rates apply to all subscribers outside Europe. Euro rates apply to all subscribers in Europe, except the UK and the Republic of Ireland where the pound sterling price applies. If you are unsure which rate applies to you please contact Customer Services in the UK. All subscriptions are payable in advance and all rates include postage. Journals are sent by air to the USA, Canada, Mexico, India, Japan and Australasia. Subscriptions are entered on an annual basis, i.e. January to December. Payment may be made by sterling cheque, dollar cheque, euro cheque, international money order, National Giro or credit cards (Amex, Visa and Mastercard).

Back Issues: Taylor & Francis retains a three year back issue stock of journals. Older volumes are held by our official stockists to whom all orders and enquiries should be addressed:
Periodicals Service Company, 11 Main Street, Germantown, NY 12526, USA. Tel: +1 518 537 4700; fax: +1 518 537 5899; email: psc@periodicals.com.

Copyright © 2010 The Textile Institute. All rights reserved. No part of this publication may be reproduced, stored, transmitted, or disseminated, in any form, or by any means, without prior written permission from Taylor & Francis, to whom all requests to reproduce copyright material should be directed, in writing.

Disclaimer: Taylor & Francis makes every effort to ensure the accuracy of all the information (the "Content") contained in its publications. However, Taylor & Francis and its agents and licensors make no representations or warranties whatsoever as to the accuracy, completeness or suitability for any purpose of the Content and disclaim all such representations and warranties whether express or implied to the maximum extent permitted by law. Any views expressed in this publication are the views of the authors and are not the views of Taylor & Francis and The Textile Institute.

Taylor & Francis grants authorisation for individuals to photocopy copyright material for private research use, on the sole basis that requests for such use are referred directly to the requestor's local Reproduction Rights Organisation (RRO). The copyright fee is £21/US$39/€30 exclusive of any charge or fee levied. In order to contact your local RRO, please contact International Federation of Reproduction Rights Organisations (IFRRO), rue du Prince Royal, 87, B-1050 Brussels, Belgium; email: iffro@skynet.be; Copyright Clearance Center Inc., 222 Rosewood Drive, Danvers, MA 01923, USA; email: info@copyright.com; or Copyright Licensing Agency, 90 Tottenham Court Road, London, W1P 0LP, UK; email: cla@cla.co.uk. This authorisation does not extend to any other kind of copying, by any means, in any form, for any purpose other than private research use.

The 2010 US Institutional subscription price is $335. Periodical postage paid at Jamaica, NY and additional mailing offices. **US Postmaster:** Send address changes to TTPR, c/o Odyssey Press, Inc., PO Box 7307, Gonic NH 03839, Address Service Requested.

Subscription records are maintained at Taylor & Francis Group, 4 Park Square, Milton Park, Abingdon, OX14 4RN, United Kingdom.

For more information on Taylor & Francis' journal publishing programme, please visit our website: www.tandf.co.uk/journals.

CONTENTS

Textile Progress
Vol. 42, No. 3, September 2010, 181–226

Geotextiles: production, properties and performance

Amit Rawal[a]*, Tahir Shah[b] and Subhash Anand[b]

[a]*Department of Textile Technology, Indian Institute of Technology, New Delhi, India;*
[b]*Institute for Materials Research and Innovation, University of Bolton, Bolton, United Kingdom*

(*Received 16 November 2009; final version received 25 November 2009*)

The monograph critically reviews most commonly used geotextile structures, their properties and performance characteristics. In general, both natural and synthetic fibres are used for the production of geotextiles, and the advantages and disadvantages of each type of fibre are discussed for various applications of geotextiles. The important functions of geotextiles, i.e. filtration, drainage, separation and reinforcement have been identified and have been related to several properties and major applications of geotextiles. Various geotextile properties, namely mechanical, hydraulic and chemical and their test methods have been critically discussed. A process–structure–property relationship for most commonly used geotextiles is also analysed. Furthermore, the design of a geotextile is of paramount importance for any civil engineering application. Thus, the design criteria for various functions of geotextiles have been addressed. Subsequently, the durability characteristics of geotextile have been introduced for analysing the performance over its lifetime.

Keywords: geotextile; mechanical; chemical; hydraulic; design; durability

1. Introduction

The word 'Geotextiles' is a combination of two words. The word 'Geo' comes from the Greek word, meaning 'Earth', and textiles. Geotextile has been defined in *Textile Terms and Definitions*, published by The Textile Institute, as 'Any permeable textile material used for filtration, drainage, separation, reinforcement and stabilisation purposes as an integral part of civil engineering structures of earth, rock or other constructional materials' [1]. Another definition of geotextiles is, 'Geotextiles are permeable textiles used in conjunction with soils or rock as an integral part of a manmade project' [2].

1.1. Historical development

Geotextiles can be made from either natural or synthetic fibres and can be utilised for both short- and long-term applications. The exploitation of the use of natural fibres in construction can be traced back to the fifth and fourth millennia BC as described in the Bible (Exodus chapter 5, verse 6–9), wherein dwellings were formed from mud/clay bricks reinforced with reed or straw. Two of the earliest surviving examples of materials strengthened by natural fibres are the Ziggurat in the ancient city of Dur-Kurigatzu (now known as Agar-Quf) and the Great Wall of China [3]. The Babylonians 3000 years ago constructed this Ziggurat using reeds in the form of woven mats and plaited ropes as reinforcements.

---

*Corresponding author. Email: amitrawal77@hotmail.com; amitrawal77@yahoo.com

ISSN 0040-5167 print/ISSN 1754-2278 online
© 2010 The Textile Institute
DOI: 10.1080/00405160903509803
http://www.informaworld.com

The Great Wall of China, completed circa 200 BC, utilised tamarisk branches to reinforce mixtures of clay and gravel [4].

Woven cotton fabrics were used as an early form of geotextile/geomembrane in a series of road construction field tests started in 1926 by the South Carolina Highways Department. As the fabric was covered with hot asphalt during construction, its function was more akin to that of a geomembrane than of a geotextile in this early application [5]. The basic difference between the geotextile and geomembrane is that the former conducts and the latter restricts the fluid flow [6]. The use of synthetic fibre-based geotextiles in ground engineering started to develop in the late 1950s, the earliest two applications being a permeable woven fabric employed underneath concrete block revetments for erosion control in Florida and in the Netherlands in 1956, where Dutch engineers commenced testing geotextiles formed from handwoven nylon strips, for the "Delta Works Schemes" [3].

The development of the first nonwoven geotextile in 1968 by the Rhone-Poulenc company in France widened the range of geotextiles and their applications. This was a comparatively thick needlepunched nonwoven made from polyester and was developed into the *Bidim* range of geotextiles. It was first used in dam construction at Valeros Dam in France during 1970. At about the same time Imperial Chemical Industries PLC (ICI) started to develop the *Terram* range of nonwoven geotextiles, which were quite different from the *Bidim* range, being a thinner heat-bonded material. The first application of this material was used as a separator between the underlying soil and the imported aggregate in the construction of temporary roads, for which the first organised trials took place in 1972.

In 1971, three other lines of geotextile application appeared, namely the first fin drains, the first woven geotextile basal reinforcement beneath embankments and the first geotextile reinforced soil wall. Initially, the fin drain concept was developed at the University of Connecticut in 1969, under contract to the Connecticut Department of Transportation. It was required to act as an easy-to-install drain system that was immune to construction errors. The concept of using woven geotextile sheets as basal reinforcement in embankments was very similar to the Japanese use of geotextile nets, except that the separation was enhanced by using continuous sheets [2].

For a fuller account of development of geotextiles and geosynthetics, readers are referred to John [2].

1.2. Functions of geotextiles

The functions of geotextiles are broadly classified into four categories as discussed below:

- *Filtration* – A geotextile acts as a filter (i) when, it comes in contact with the liquid carrying fine particles, therefore it stops the majority of particles, whilst liquid, passing through; (ii) when, placed in contact with the soil, it allows seeping of water retaining most of the soil particles being carried away by water current. In general, the volumetric flow of water passing through the geotextile determines its suitability for filtration function.
- *Drainage* – A geotextile acts as a drain when it collects and redirects liquid or gas towards the outlet, i.e. the transmission of fluid is in the direction of in-plane flow of fabric without any loss of soil particles [7]. Any geotextile material exhibiting good filtration and permittivity characteristics can be used in drainage applications [8].
- *Separation* – A geotextile acts as a separator when placed between fine soil and coarse material. It segregates the materials and prevents mixing, especially at applied

loads. In other words, this function of geotextile is simply to prevent the gravel from penetrating and mixing with the subgrade [9].

- *Reinforcement* – A geotextile acts as reinforcement when the stability of the weak subgrade or soil is complemented by higher tensile strength of fabric. The principle of using geotextiles as reinforcement is to introduce the geotextiles into the soil structure that increase the cohesion between the grains [10]. This modifies the transmission of the load and the resulting composite is able to sustain higher loads. The forces exerted on the structure as a result of various loads are transferred into tensile stresses, which further influences other mechanical properties, such as puncture resistance [11]. The reinforcement is a complex phenomenon and results from the combined behaviour of soil–geotextile interactions [12–14].

In addition, other functions of geotextiles have been identified by Giroud [6] and the definitions of these functions are given below.

- *Surfacing* – A geotextile acts as a surfacing when a smooth and flat ground surface is required and preventing the soil particles to be removed from the soil surface.
- *Solid barrier* – A geotextile acts as a solid barrier when it prevents or ceases the motion of solids.
- *Container* – A geotextile acts as a container when it holds or protects the materials such as sand, rocks, fresh concrete etc.
- *Tensioned membrane* – A geotextile acts as a tensioned membrane when it is sandwiched between two materials having different pressures. The principle of using a geotextile is to even out the pressure difference by balancing with the tension of the geotextile.
- *Tie* – A geotextile acts as a tie when it joins various pieces of a structure that is capable of moving apart.
- *Slip surface* – A geotextile placed between two materials by minimising the frictional characteristics of the structure.
- *Absorber* – A geotextile acts as an absorber when it shares the stresses and strains transmitted to the material that is required to be protected.

1.3. Fibres used in geotextiles

In general, synthetic and natural fibres are used in the production of geotextiles. Natural fibres can be of plant, animal and mineral origin and vast quantities of these are available worldwide. Natural fibres offer high strength, high modulus, low breaking extension and low elasticity. Figure 1 illustrates the stress–strain properties of various natural fibre yarns [15]. It should be noted that natural fibre yarns and fabrics possess low levels of creep during use. Mineral fibres are brittle and lack strength and flexibility. Since tensile strength is an important property for a geotextile, especially for reinforcement applications, plant fibres show the greatest potential for use in geotextiles. Some of the plant fibres that can be used in geotextile manufacture are jute, sisal, flax, hemp, abaca, ramie and coir. The natural fibre-based geotextiles are biodegradable, therefore these materials could be specifically used for temporary functions. In summary, the main advantages of using natural fibres in geotextiles are low cost, robustness, strength/durability, availability, good drapeability and biodegradability/environment friendliness. Table 1 shows the properties of the most important natural fibres that are used in the production of geotextiles.

Synthetic fibres are the main raw materials for the manufacture of all types of geotextiles. Geosynthetics is the universally accepted name for synthetics used in intimate association

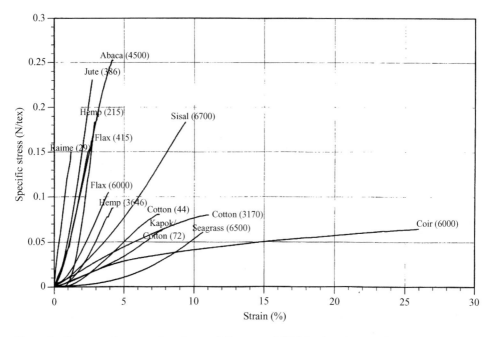

Figure 1. Stress–strain properties of natural fibre yarns [15]. Numbers in parentheses represent yarn tex. Reprinted from S.C. Anand, Indian J. Fibre Text. Res. 33 (2008) pp. 339–344, with permission of NISCAIR.

with geotechnical materials and can be classified as shown in Figure 2. There are four predominant polymer families used as raw materials for geosynthetics, i.e. polypropylene, polyester, polyamide and polyethylene. Table 2 illustrates the comparative properties of the above-mentioned polymer families used in geosynthetics. It should be noted that within these polymer groups there are many subgroups and variants, each with their own set of characteristics. The general comparisons shown in Table 2 are therefore very broad and refer to the polymer forms most likely to be encountered in geotextiles. Nevertheless, polypropylene has been the most widely used polymer for the production of geotextiles due to its low cost, acceptable tensile properties and chemical inertness. Polypropylene has low density, which results in very low cost per unit volume. The disadvantages of polypropylene are its sensitivity to ultraviolet (UV) radiation and high temperature and poor creep and mineral oil resistance. Thus, polypropylene-based geotextiles should be used under suitable installation and environmental conditions. Similarly, another important

Table 1. Properties of natural fibres.

Property	Type of fibre					
	Flax	Jute	Hemp	Sisal	Abaca	Coir
Fibre length (mm)	200–1400	1500–3600	1000–3000	600–1000	1000–2000	150–350
Fibre diameter (mm)	0.04–0.62	0.03–0.14	0.16	0.1–0.46	0.01–0.28	0.1–0.45
Fibre linear density (dtex)	2–20	14–20	3–22	10–450	40–440	–
Fibre tenacity (Ntex^{-1})	0.54–0.57	0.41–0.52	0.47–0.60	0.36–0.44	0.35–0.67	0.18

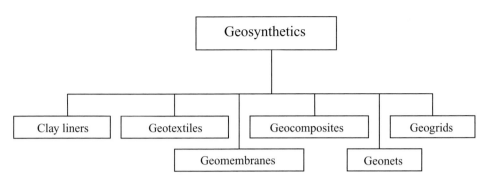

Figure 2. Classification of geosynthetics.

synthetic polymer used in the manufacture of geotextiles is polyethylene terephthalate (PET), commonly referred to as polyester. It exhibits superior creep resistance and tenacity values and is used in applications where the geotextile is subjected to high stresses and elevated temperatures. The main disadvantage of polyester is susceptibility to hydrolytic degradation in soils exceeding pH 10. Other synthetic fibres such as polyamides (nylon 6,6 and nylon 6) are used in small quantities as conventional geotextiles. It should also be borne in mind that although the choice of polymer has a great influence on geotextiles' strength and creep properties, other factors such as fabric structure, finishing treatments applied to the fabric and the confining effect of any surrounding soil could also influence their properties [2].

1.3.1. Synthetic vs. natural fibres for geotextile applications

In general, synthetic fibres, such as polypropylene, polyester, polyethylene, polyamide etc., have dominated the geotextile industry, although the usefulness of natural fibres should not be ignored, because the latter are environment friendly, less costly, easily available and ecologically compatible as they are degraded within the soil [16]. However, the fibre requirements are heavily dependent on the envisaged applications. Natural fibres are mainly

Table 2.   Comparative properties of general polymer families.

Comparative properties	Polymer group			
	Polyester	Polyamide	Polypropylene	Polyethylene
Strength	1	2	3	3
Elastic modulus	1	2	3	3
Strain at failure	2	2	1	1
Creep	3	2	1	1
Unit weight	1	2	3	3
Resistance to				
UV light				
Stabilized	1	2	1	1
Unstabilized	1	2	2	3
Alkalis	3	1	1	1
Fungus, vermin, insects	2	2	2	1

1 = High
2 = Moderate
3 = Low

employed for short-term reinforcements and erosion-control applications of geotextiles. For instance, in slope stabilisation, natural fibres including jute, hemp or coir are required for a relatively short period of time in order to establish the root structures after which the geotextile is required to be decomposed for visual aesthetic reasons. Several researchers have demonstrated the use of natural fibres including jute, coir, wood, flax and bamboo in various applications of geotextiles such as soil erosion control, vertical drains, road bases, bank protection and slope stabilisation [16–26]. In addition, Ranganathan [27] has revealed the potential of jute-based geotextiles for new products and applications such as supersod, temporary haul roads, reinforcement fabric in highway construction, wick drains etc. because these geotextiles have high water absorption and moisture uptake, which makes them ideal materials for such applications. Similarly, the employment of a coir-based geotextile has shown a remarkable improvement in the vegetal growth [16]. However, the coir geotextile is degraded due to the microbial action in the soil in addition to the effect of rain and sun. Lekha [16] has observed that coir net retained only 22% of its initial tensile strength at the end of seven months after it was deployed in the soil. Similar strength loss in coir netting was reported by Balan and Venkatappa Rao [28]. Thus, in applications where natural fibres are exposed to microbiological agents and solar radiation, such fibres are expected to have reduced effectiveness [29]. The effect of solar radiation is not limited to natural fibres but synthetic fibres, such as polypropylene, also have a poor resistance to UV radiation. Similarly, nylon with a higher tensile strength than polyester or polypropylene tends to be degraded by weathering [30]. However, nylon can resist at least twice the level of abrasion in comparison to polyester or polypropylene fibre [31], but polyester has higher abrasion resistance on exposure to UV light, whereas polypropylene fibre has a superior resistance to fatigue-flexing. In roofing applications whereby the fabric is placed under higher tensile and flexural stresses and also subjected to abrasion or bursting stresses, polyester fibre with higher tenacity and lower elongation is more suitable. Polyester fibre is also least affected by acidic conditions or changes in temperature that occur due to seasonal variations. It has also been suggested that polyester fibre can be used for tidal-barrage protective devices for the same reasons of resistance to solar radiation and mechanical stresses, in addition to resistance to salt solutions [32]. On the other hand, polypropylene fibre with lower density leading to better buoyancy properties is more appropriate for tidal barrages, which are frequently subjected to battering [25]. It can be summarised that the choice of the fibre is realised due to its degradative conditions and needs to be judiciously selected for the desired area of application of geotextiles.

1.3.2. Major geotextile applications

The following are the specific areas of applications where both natural and synthetic fibre-based geotextiles have been used:

1. River bank protection.
2. Seabed protection.
3. Sea coastal protection.
4. Coastal protection/sea defences.
5. Drainage.
6. Perpendicular versus in-plane water flow applications.
7. Reservoirs and lakes.
8. Concrete mattresses.
9. Vertical screens.
10. Geobags.

Table 3. Geotextile applications and functions.

	Geotextile function			
Application	Reinforcement	Separation	Filtration	Drainage
Roads	P	T	T	S
Railroads	S	P	S	P
Drainage	T	S	S	P
Steep slopes	P	T	T	S
Landfills	S	P	P	S
Walls	P	T	T	T
Soil reinforcement	P	T	S	S
Land reclamation	S	S	P	S
Marine causeways	S	T	S	P
River protection	S	T	P	S

Notes: P = primary function; S = secondary function; T = tertiary function.

11. Reclaimed land.
12. Anchoring.
13. Paved and unpaved road subgrade separation.
14. Embankments.
15. Reinforced earth.
16. Reflective cracking.
17. Railways.
18. Landfills.
19. Slit fences.

Table 3 shows the functions of geotextiles and their relationship with the required applications [33].

1.4. Production of geotextiles

The majority of geotextiles are produced from classical or conventional fabric-production techniques. Giroud [6] has broadly classified the manufacturing processes for the production of geotextiles in two classes, i.e. classical and special geotextiles. In classical geotextiles, typical products of textile industry, such as woven, knitted, nonwoven fabrics, etc., are considered, whereas special geotextiles are similar in appearance to that of classical geotextiles but are not the products of textile industry, i.e. webbing, mats and nets. A typical classification of production of geotextiles is shown in Figure 3 [6]. Classical geotextiles are produced in two steps, i.e. production of fibres, filaments, slit films (tapes) and yarns and converting these constituent materials into a fabric. The constituent materials required for making a fabric are produced using various techniques as discussed below) [6]:

Filaments: In a typical melt extrusion process, the filaments are produced by extruding molten polymer through spinnerets or dies. Subsequently, the filaments are drawn so that molecular orientation along the filament is improved resulting in high tensile strength and modulus. Furthermore, when there are numerous filaments being extruded through the spinneret, it is known as *multifilament* yarn.

Short (staple) fibres: Filaments are cut into short lengths ranging from 2 to 10 cm and when these fibres are twisted together, they form a yarn.

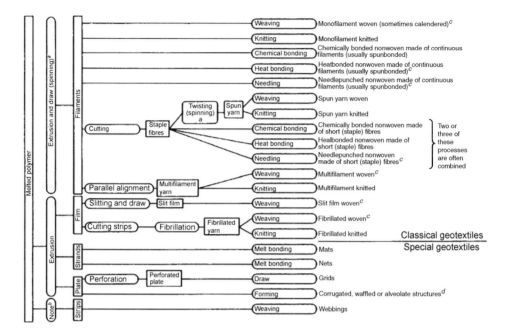

Figure 3.   Production of geotextiles [6]. Products are in lower case letters, processes are in capital letters. All special geotextiles are currently used. Reprinted from J.P. Giroud, Geotext. Geomembranes 1 (1984b) pp. 5–40, with permission of Elsevier.

Notes: [a]The word spinning has two meanings, extrusion through a spinneret to make a filament, and fabrication of yarns from staple fibres. [b]Strips can be made using any appropriate process such as extrusion, calendering, weaving, yarn, fabric coating etc. [c]Some classical geotextiles (those indicated) are used more than others. [d]Corrugated, waffled or alveolate structures are generally not used alone, they are used to make composite geotextiles.

Slit films: The films are produced through a melt extrusion process using slit dies, which are subsequently slitted with sharp blades. These films can be further fibrillated and broken into fibrous strands known as *fibrillated yarn*.

The above linear elements, namely filaments, fibres, slit films or yarns are converted into different types of classical geotextiles as briefly described below.

Woven geotextiles: A woven fabric consists of two sets of interlaced filaments or yarns that are generally orthogonal to each other. The weave design or pattern is determined by the manner in which yarns or filaments are interlaced. Filaments or yarns in the longitudinal and transverse directions are known as warp and weft, respectively. Monofilament and slit-woven geotextiles are thinner in comparison to multifilament, spun and fibrillated woven geotextiles.

Nonwoven geotextiles: Nonwoven fabrics are defined as a sheet, web or batt of directionally or randomly oriented fibres/filaments, bonded either by friction and/or cohesion and/or adhesion. In general, nonwoven fabric formation is a two-step process, i.e. web formation (laying up the fibres with certain orientation characteristics) and bonding the fibres by mechanical, thermal or chemical means. This two-step process forms the basis of classification of nonwoven structures, i.e. carded, air-laid, spunbonded, meltblown, needlepunched, hydroentangled, adhesive bonded, thermal bonded, stitch bonded etc. Some of the important processes that are used for the production of nonwoven geotextiles are discussed below.

- Spunbonding: This process includes filament extrusion, drawing, lay down and bonding. The first two steps are basically adapted from a typical melt extrusion process. The latter steps involve the deposition of filaments on to a conveyor belt in more or less random fashion. It should be noted that the spunbonded nonwovens are generally self-bonded but they can be subsequently bonded additionally by thermal, chemical or mechanical means for further improvement in the tensile properties.
- Chemical bonding: A binder such as glue, rubber, casein, latex, cellulose derivative or a synthetic resin is added to fix together filaments or short fibres to produce a chemically bonded nonwoven.
- Mechanical bonding: It is classified into two categories, i.e. needlepunching and hydroentanglement (also referred to as spunlacing). The difference in mechanical bonding techniques relies on the utilisation of metal needles in needlepunching, whereas high-pressure multiple rows of water jets are used to reorient and entangle a loose array of fibres into self-locking and coherent fabric structures in a hydroentanglement process [34].
- Thermal bonding: These are produced by applying the heat energy to the thermoplastic component present in fibrous web and the polymer flows by surface tension and capillary action to form the required number of bonds at crossover positions of fibres [35]. It can be mainly classified in two categories, i.e. through-air bonding and calendaring. In through-air bonding process, the fibrous web is passed in a heated air chamber for forming the heated bonds at the crossover positions of fibres. The latter process involves the passage of fibrous web through a heated pair of rollers that impart high pressure and temperature.

Knitted geotextiles: These are produced by interlocking a series of loops of filaments or staple-fibre yarns to form a planar structure. The loops in the knitted structure can be interlocked in different ways similar to that of weave design in woven fabrics.

Braided geotextiles: These are narrow rope-like structures consisting of yarns interlaced at a bias direction. The braided structure is generally tubular in nature and different designs such as diamond, regular and Hercules similar to plain, 2/2 twill and 3/3 twill incorporated in a woven structure [36].

Similarly, special geotextiles are also manufactured using a two-step process as briefly discussed below.

Webbings: These are produced from strips of moderate width and are similar to coarse woven slit film fabrics.

Mats: These are made of coarse and rigid filaments having tortuous shape similar to that of open nonwoven fabrics.

Nets: These consist of two sets of inclined coarse parallel-extruded strands and are bonded at the intersections by partially melting one of the strands. These net structures can also be produced using a melt extrusion process consisting of rotating dies through which the molten polymer is extruded [37].

In addition, composite geotextiles can be produced by combining several of the above products such as a combination of multiple layers of knitted/woven/nonwoven by means of stitching, needlepunching, thermal bonding etc. Similarly, mats/nets/plastic sheets can be sandwiched with one or two geotextiles especially for drainage applications.

1.5. Market outlook

Geotextiles are forecast to be the fastest growing sector within the market for technical textiles. The UK-based consultants David Rigby Associates had predicted the average annual growth of 4.6% for geotextiles market between 2000 and 2005 in terms of volumes of fibre consumed, accelerating to 5.3% per annum between 2005 and 2010. By comparison, the market for technical textiles as a whole was forecast to grow by 3.3% between 2000 and 2005 and by 3.8% between 2005 and 2010 [38].

The fastest growth will be expected in China and other East Asian countries. This is mainly due to the fact that (1) such countries are late entrants, which means that their geotextile markets are relatively immature; and (2) the authorities within those countries are initiating and undertaking major infrastructure projects, which require large amounts of geotextiles. Asia and other developing regions of the world are much more likely to see a sustained programme of infrastructure projects over the next decade and beyond. The Chinese "Three Gorges Project" is one such example, which has given a considerable stimulus to the use and local manufacture of geotextiles.

Within North America and Western Europe, a more dynamic area of activity – but one which is of smaller volume – is the design and development of value-added products for high technology applications. Table 4 shows the end use consumption of geotextiles selected by region.

An analysis of the market by fibre type shows that polyester and polypropylene fibres are the most commonly used polymers in the manufacture of geotextiles. In North American and Western European markets, 80% of the geotextiles are made from polypropylene. Polyester is mainly used in applications where high creep resistance and UV stability are needed [33]. However, natural fibre geotextiles are being made available in increasing amounts in Bangladesh, China, India and other developing countries which have indigenous supplies of suitable natural fibres, such as jute, sisal, coir etc.

From the fabric structure viewpoint, nonwoven geotextiles have grown at a much faster rate than their woven couterparts. More than 75% of the market utilises geotextiles made from various nonwoven fabrics, chiefly staple-fibre needlepunched and continuous-filament spunbonded nonwovens. Nonwoven materials are normally 25–30% cheaper than woven materials of a similar area density (grams per square metre) and fibre type. Warp knitted fabrics are normally 10–15% cheaper than woven fabrics.

Other significant current and future developments in this market segment are geo-composites – combinations of two or more separate materials in one product, and the development of multi-functional geotextiles. These will be able to perform a variety of

Table 4.   Growth in end use consumption of geotextiles by volume, by selected region 1995–2010 (% per annum).

	Annual average percentage change		
	1995–2000 (a)	2000–2005 (b)	2005–2010 (b)
North America	3.9	1.8	2.7
Western Europe	6.0	5.9	6.0
East Asia	8.6	7.7	8.1
World	5.4	4.6	5.3

Notes: (a) = estimates; (b) = forecasts.

functions and will be suitable for widely different applications and environmental conditions [33].

2. Properties of geotextiles

Some of the important properties of geotextiles are discussed below.

2.1. Mechanical properties of geotextiles

2.1.1. Tensile and puncture properties of geotextiles

The two commonly used geotextiles are woven and nonwoven, and the deformation of these structures under uni- or biaxial loading is significantly different [10]. In general, nonwovens are less stiff than the woven material but have higher extensibility, which is *ideal* for certain geotextile applications. However, the production technique of nonwovens or type of weave plays an important role in fulfilling the requirements of geotextiles. Typical load-extension curves of woven and nonwoven geotextiles are shown in Figure 4 [10].

Gautier, Kocher, and Drean [39] have analysed the tensile behaviour of nonwoven geotextiles using optical strain-mapping techniques along with *infrared thermography*. They have tried to establish that needlepunched nonwovens are more anisotropic in nature in comparison to thermal or heat-bonded nonwovens. However, no information about the structure of nonwovens was provided which can help in drawing such conclusions. The tensile properties of a geotextile are generally determined through strip or grab test methods [10], but these should be measured in a manner which simulates the operational conditions of geotextiles, i.e. under confined stresses. In the past, many researchers have suggested that the tensile characteristics of geotextiles should be determined under confined conditions [40–45]. Andrawes, McGown, and Kabir [46] have reported the uniaxial tests on woven and nonwoven geotextiles by investigating the effect of width-to-length ratio of the specimens, test temperature, pressure and strain rate. It has been demonstrated that woven geotextiles are more susceptible to changes in the strain rate and temperature than nonwoven geotextiles. Nonwovens undergo structural realignment under uniaxial loading, whereas wovens have straight yarns, which are highly sensitive to temperature and strain rate. Here, the intrinsic properties of polymer play a key role in determining the deformation behaviour of woven geotextiles. Furthermore, it was also found that the shapes of stress–strain curves are significantly affected at higher confined stresses and cannot be simulated

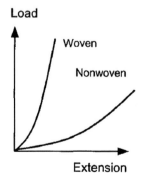

Figure 4. Load-extension curve of woven and nonwoven geotextiles [10]. Reprinted from Y. Wang, J. Ind. Text. 30 (2001) pp. 289–302, with permission of SAGE Publications Ltd.

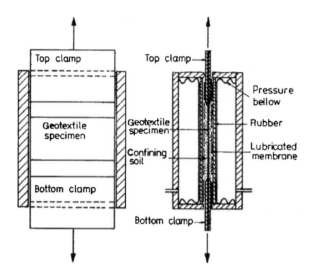

Figure 5. An apparatus for in-soil testing [46]. Reprinted from K.Z. Andrawes, A. McGown and M.H. Kabir, Geotext. Geomembranes 1 (1984) pp. 41–56, with permission of Elsevier.

with a wider specimen size [47]. Therefore, an apparatus (shown in Figure 5) consisting of metal box with two bellows for applying confined pressure through a soil layer was built but thin clamps were used, which limited its usage for general testing. Subsequently, a stand-alone pressure chamber using ordinary grips was developed for a confined tensile test [10]. Here a composite specimen (see Figure 6) consisting of a glued section at the ends and unglued region in the middle portion were prepared and tested with and without confining pressure. Hence, the stress–strain response of the glued region under confined pressure can be determined by testing a fully glued sample without confined stresses. However, the properties of glue are required to be *known* in order to obtain accurate results. Nevertheless, one of the most significant findings obtained from these studies of tensile tests under confined conditions is that there is a significant increase in the stiffness and strength of the geotextiles in comparison to the unconfined conditions. However, it is quite contentious as these in-soil tests do not simulate the *real* conditions of a geotextile in a reinforced structure as it overcomes the frictional resistance against the stationary soil before the tensile load

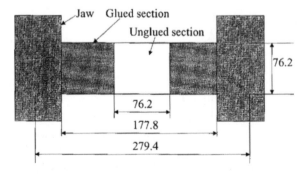

Figure 6. Composite specimen consisting of glued and unglued regions [10]. Dimensions are in mm. Reprinted from Y. Wang, J. Ind. Text. 30 (2001) pp. 289–302, with permission of SAGE Publications Ltd.

within the structure is mobilised and this results in load corresponding to the combined effects of the frictional force and the stress confinement [41]. In geotextile-reinforced structures, slippage at the soil–geotextile interface does not occur until a failure state is approached, which tends to overestimate the strength and stiffness of a geotextile.

During the application of uniaxial tensile load on a geotextile, the center region tends to decrease in size with increase in longitudinal strain resulting in a parabolic shape for the sides of the deformed specimen, also known as *necking* [48]. Theoretically, it is assumed that lateral strain is uniform over the specimen length but is zero at the two ends of the specimen due to clamping and is maximum at the centre region of the specimen and hence Poisson's ratio (ratio of lateral contraction to the longitudinal strain) tends to be overestimated. Nevertheless, Poisson's ratio (v) can be computed on the assumption that the ratio of volume (V) of geotextile corresponding to strain (ε) to volume (V_0) of geotextile at the beginning of tensile test is a linear function of applied strain (ε), as shown in Equation (1) [48].

$$v = \frac{1}{\varepsilon}\left(1 - \sqrt{\frac{1 + \varepsilon(1 - 2v_0)}{1 + \varepsilon}}\right), \tag{1}$$

where v_0 is the Poisson's ratio for zero strain.

Furthermore, the other mechanical properties, such as puncture resistance, are related to tensile properties of geotextiles as shown below [11,49].

$$T_f = \frac{F_p}{2\pi r}, \tag{2}$$

where T_f is the tensile force per unit width, r is the radius of plunger and F_p is the puncture force.

The above equation revealing the relationship between puncture force and tensile strength has been modified for axisymmetric condition [50–52].

$$T_f = \frac{F_p}{2\pi r \sin\alpha} \tag{3}$$

and

$$\sin\alpha = \frac{y}{\sqrt{y^2 + 50^2}}, \tag{4}$$

where α is the angle between the geotextile plane and initial horizontal plane, and y is the vertical displacement of puncture rod as shown in Figure 7 [52].

Bergado et al. [52] have suggested that loading from moving vehicle wheels is approximated to an axisymmetric loading rather than in-plane loading. However, it has been pointed out that geotextile plane is slightly curved between the edge of puncture rod and the circular clamps. In addition, the tensile strength of nonwoven geotextile is found to be independent of strain rate, matching with previous results of Maneecharoen [53] and McGown et al. [50]. Nevertheless, Resl and Warner [54] have also demonstrated that needlepunched nonwoven geotextiles under axisymmetric loading can significantly increase the bearing capacity of soft grades.

It is interesting to note that tensile strength can be determined from puncture resistance based on Equations (2)–(4), and it has been reported that the measured wide width tensile

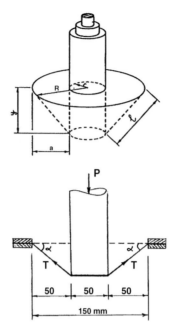

Figure 7. Geometry to interpret an axisymmetric tension–strain of geotextile in CBR puncture test [52]. Reprinted from D.T. Bergado, S. Youwai, C.N. Hai and P. Voottipruex, Geotext. Geomembranes 19 (2001) pp. 299–328, with permission of Elsevier.

strength of woven geotextiles was found to be higher than the strength computed from puncture tests and the vice versa results were obtained for nonwoven geotextiles [55]. Puncture resistance has also been related to fabric area density and thickness [56,57]. Similar results have been established by Rawal, Anand, and Shah [58] demonstrating the effect of puncture resistance on fabric area density and fabric thickness. According to Ghosh [11], a geotextile is subjected to concentrated forces normal to the plane while the fabric is pretensioned and the distribution and magnitude of these forces result in the puncture failure of geotextiles. Their results indicated that lower strain in puncture occurred if the geotextile is prestrained. It was also concluded that biaxial failure simulated puncture failure more appropriately than uniaxial tests. Similarly, Giroud [59] had pointed out that highly extensible geotextiles do not sustain damage, although subsurface is uneven and consists of irregularities. The puncture properties of geotextiles are also affected by soil parameters, which include shear strength, angularity and bluntness [60,61].

2.1.2. Frictional behaviour of geotextiles

Soil particles have good shear and compressional resistance but are quite weak against tension. Thus, a tensile element such as straw or galvanised steel strips was initially used within the soil for improving its strength [62,63]. Over the years, geotextiles have replaced straws and strips as they have low stiffness and are relatively more compatible with the soil in terms of deformability. Hence, geotextiles are used as tensile elements for reinforcing the soil, transfer most of the shear stresses from soil to reinforcement by surface friction and any remaining stress by interlock [2,64]. Alternatively, another mechanism of soil–geotextiles behaviour was proposed indicating the passive resistance of soil against

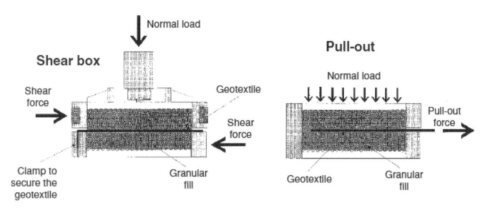

Figure 8. Test methods to determine frictional resistance of a geotextile [4]. Reprinted from A.R. Horrocks and S.C. Anand (eds.), *Handbook of Technical Textiles*, Woodhead, Cambridge, UK, 2000, with permission of Woodhead Publishing Limited, UK.

reinforcing elements (geotextiles) placed normal to the direction of shear displacement [65]. In general, interfacial shear resistance between soil and geotextiles is generated by friction bonding and interlocking. The former is governed by the surface roughness of geotextiles, whereas the latter is dictated by the pore size and deformation capability. The efficiency of geotextiles in developing shear resistance at the soil–geotextile interface is known by contact efficiency or coefficient of interaction (α), which is defined as the ratio of the tangent of the friction angle (δ) for the soil–geotextile to the tangent of the friction angle of the soil alone (ϕ), i.e. $\tan\delta/\tan\phi$.

The soil–geotextile interaction can be classified into two major forms, i.e. shearing (bond) and pull-out (anchorage) interactions [66,67]. The former deals with the conditions of soil sliding over a geotextile, whereas the latter reveals the conditions related with the slippage of geotextile in reinforced soil structures. These forms of soil–geotextile interaction can be assessed using direct shear and pull-out tests, as illustrated in Figure 8 [4]. In the direct shear box test, the soil is strained against the fabric, whereas in the pull-out test, strain is applied to the fabric such that shear resistance along the fabric is mobilised with respect to the applied load and the fabric extensibility [4]. In general, the soil–geotextile shearing test can be easily carried out but the deformation of geotextiles is not mobilised, whereas the pullout is difficult to interpret because of the complex interaction behaviour at the interface [66,68,69]. Nevertheless, the pull-out test is more closely related to the failure mechanism of reinforced soil structure and can be computed as shown in Equation (5) [70].

$$P = S + F, \tag{5}$$

where
P = pull-out force,
S = shearing force,
F = frictional force.

It has been pointed out by several researchers that the frictional resistances developed due to pull-out forces are not uniformly distributed over the geotextile surface [69,71]. However, Kharchafi and Dysli [72] studied the movement of soil and fabric during the pull-out test using X-ray radiography. They pointed out that there is a distinct failure of two geotextiles having different stiffness, i.e. failure in less stiff geotextile occurs by the rupture of the geotextile itself, whereas a stronger geotextile occurs by the lack of adherence. Furthermore, the stiffer geotextile yielded linear strain distribution and is distributed over a greater length. Surprisingly, the soil–geotextile friction angle was found to be independent of geotextile stiffness. It was also found that the stiff geotextile distributes the friction over a larger surface area and hence improved soil–geotextile adherence can be obtained. The friction between the soil and geotextile depends on various parameters including geotextile properties (type, constituent fibres and their properties, fibre orientation, pore dimensions, mass per unit area, thickness, extensibility, surface roughness, anchor length), soil properties (size, shape, density, conditions), loading conditions (normal stress component), rate of deformation and method of testing [12,14,65,68,70,73–76].

On the other hand, several researchers have studied the interaction between soil and the geotextile based on shear test necessary for obtaining bond parameters for design purposes [66,77–79]. The shear strength at a soil–geotextile interface is a function of the stress normal to the interface. The interface shear strength and normal stress are linearly related in the following form:

$$\tau = a + \sigma \tan \delta, \qquad (6)$$

where τ is the soil–geotextile or geotextile–geotextile interface shear strength, a is the interface adhesion, σ is the stress normal to the interface and δ is the interface friction angle.

Giroud, Darrase, and Bachus [80] have pointed out that a linear relationship between the interface shear strength and stress normal to the interface may lead to errors on safety factor of the slope, especially at lower normal stresses. Thus, they proposed the hyperbolic relationships as shown in the following equation:

$$\tau = a_\infty + \sigma \tan \delta_\infty, \qquad (7)$$

where a_∞ is the pseudo-adhesion, i.e. intercept of the asymptote with τ axis and δ_∞ is the interface friction angle for very large values of σ (slope of the asymptote). Equation (7) illustrates an empirical relationship containing number of constants that do not have physical meaning [81].

Bouazza and Djafer-Khodja [77] have shown that the friction coefficient decreased with the increasing stress due to interface adhesion and the transmissivity of geotextile decreased at higher normal stresses causing the interaction between the geotextile and peat to become undrained [82]. According to Athanasopoulos [81], the shear tests were conducted for woven and nonwoven geotextiles using cohesive soils having water content much higher than the optimum value, obtained from compaction tests. It has been demonstrated that woven geotextiles with higher tensile strength were not able to provide substantial strength increase to soils due to negligible transmissivity. However, nonwovens with reasonable strength and possessing the capability of in-plane flow of water were successfully employed in reinforcing near-saturated cohesive soils. The shear properties of soil–geotextile composite have also been studied using torsional ring and inclined board test methods [83,84]. It has been concluded that direct shear test has yielded the higher sand–geotextile peak friction angle than the ring shear test method and similarly higher interface shear strengths than those of inclined board test method.

2.1.3. *Compressional behaviour of geotextiles*

Geotextiles are subjected to static or dynamic compressional load in road and railway applications. For example, people or vehicles passing on the road can produce a dynamic effect on geotextiles. The geotextile is expected to absorb energy and should ideally maintain its structural characteristics after recovery.

A pioneering work was carried out by van Wyk [85] on the compressibility of random fibrous assemblies, in which the compressional behaviour of fibrous assemblies was expressed in terms of bending strains of constituent fibres neglecting the twisting, shearing, slippage and extension. Nevertheless, the theory calculated the distance between the contacts in a random assembly of fibres. Subsequently, the theory was modified by taking into account the change in the orientation of fibres during compression [86]. However, the results for calculating the number of fibre contacts were not validated experimentally. The compressional deformation of "general" fibrous assemblies was also predicted based upon these theories [87]. The compressional behaviour of needlepunched and spunbonded nonwovens has been proposed by Kothari and Das [88] using the theories proposed by Carnaby and Pan [87] and Lee and Lee [89]. The authors have claimed that these theories can be applied to any kind of textile material ranging from sliver, yarn to nonwovens. More recently, Rawal [90] has successfully applied the proven theories of compression to predict the compressional behaviour of thermally bonded nonwovens. The effect of fibre orientation characteristics and volume fraction on the compressional behaviour of thermally bonded nonwovens has also been investigated.

Kothari and Das [91–93] described the compressional behaviour of various nonwoven structures under dynamic loading in terms of two parameters, i.e. compressional (α) and recovery (β). These compressional and recovery parameters are dimensionless constants indicating the compression and recovery behaviour of fabrics. These parameters were deduced by formulating the relationship between pressure and thickness of the nonwoven fabrics in the uncompressed and compressed states, as shown in the Equations (8) and (9).

$$\alpha = \frac{(T_0 - T_f)/T_0}{\ln\left(\frac{P_f}{P_0}\right)}, \tag{8}$$

$$\beta = \frac{\ln(T_f/T_0)}{\ln(P_0/P_f)}, \tag{9}$$

where T_0 and T_f are the initial and final thicknesses and P_0 and P_f are the initial and final pressures, respectively. It was shown that these compressional and recovery parameters stabilise after five or six cycles of loading and there is no change in thickness. Accordingly, the geometrical parameter, namely pore size or apparent opening size, which is dependent upon thickness, would not be changed. The relationship between thickness and pore size is shown in Equation (10) [94].

$$\frac{O + d_f}{O' + d_f} = \sqrt{\frac{T}{T'}}, \tag{10}$$

where d_f is the filament diameter, O and O' is the average opening or pore size corresponding to the thicknesses T and T', respectively.

Furthermore, Sengupta, Ray, and Majumdar [95] have also revealed similar results of variation in thickness with cyclic loading. It was reported that the thickness loss increases with the increase in the cycles of dynamic loading up to a certain limit and subsequently there is no change in the thickness. More recently, Rawal [96] has established that needlepunched nonwoven structures produced under various processing conditions yielded different compressional characteristics, whereas in the past the literature has cited only one type of compressional behaviour, i.e. the compressional and recovery parameters decrease with the increase in loading cycles and subsequently become constant at a higher number of loading cycles [91,93].

2.1.4. Creep behaviour of geotextiles

When the stress is applied constantly, there is an increase in the extension with time in the materials, this phenomenon is known as *creep*. According to Hoedt [97], elongation of a fabric plays an important role in performance characteristics of geotextiles. For instance, in separation function of geotextile, ideally the fabric should be capable of separating soil layers; however, as the fabric is allowed to elongate, it rarely contributes in reducing soil deformations. Similarly, higher elongation can result in higher pore size, which may affect the filtration characteristics of geotextiles [98]. It has been pointed out in the literature that material properties, construction of geotextiles, loading conditions, temperature and time are important factors in controlling the creep behaviour of geotextiles [97]. It has been suggested that creep in woven geotextiles can be minimised by reducing the crimp in the constituent yarns. The deformation of amorphous regions in polymers also has a significant effect on the stress–strain behaviour of geotextiles and in the prolonged loading. It is one of the reasons that polyolefins, such as polyethylene and polypropylene (70–80% crystallinity), have higher creep characteristics. In general, the creep recovery strain contains four parts, i.e. instantaneous elastic, instantaneous plastic, viscoelastic and viscoplastic [97]. Schaperys' nonlinear creep equation has been modified based on arbitrary material parameters as shown in the following equation [99]:

$$\varepsilon_{(t)} = D_0^e g_0^e \sigma + D_0^p g_0^p \sigma + \int_0^t D_{\varphi - \varphi'} \frac{\partial (g\sigma)}{\partial \tau} d\tau + m\sigma^n t^l, \qquad (11)$$

where $\varepsilon_{(t)}$ is the creep, σ is the applied stress, t is the time, D_0^e is the instantaneous elastic compliance, D_0^p is the instantaneous plastic compliance, $D_{\varphi - \varphi'}$ is the creep compliance, m, n and l are the material parameters and g_0^e, g_0, g and a_σ are the stress-dependent parameters.

Other empirical models of creep particularly for geotextiles have also been proposed, which describe both the creep during initial loading and the strain recovery after unloading [100]. These models have included the effect of plastic strains that are developed in geotextiles during the loading. Furthermore, *accelerated time–temperature* procedures were used to determine the creep deformations, creep failure and residual tensile strength of geotextiles [101]. In this study, the well-known Arrhenius equation revealing the relationship between the rates of reaction and temperature along with William–Landel–Ferry (WLF) equation and Boltzmann superposition principle were combined to evaluate the *master creep curve* of geotextiles.

2.2. Hydraulic properties of geotextiles

Generally, geotextiles are porous in nature and the hydraulic properties including in-plane (transmissivity) and cross-plane permeability are significantly influenced by the voids or pores available in the geotextile, because it is well known that the flow in the medium occurs only if these pores are interconnected [102]. According to Komori and Makashima [103], the mechanics of fluid flow depend on the width of pore, which determines the quantity of the fluid flow, and on the depth of pore, which represents the free path of the fluid flow. Therefore, porous structure of geotextile is of paramount importance, specifically for filtration, drainage and separation functions of geotextiles.

2.2.1. Porous structure of geotextiles

The structure of a fibrous network in a geotextile plays a key role on the fluid flow characteristics [104]. For example, pores inside the nonwoven geotextiles are highly complex in terms of sizes, shapes and the capillary geometries and can be characterised in terms of the total pore volume (or porosity), minimum and maximum pore sizes, pore size distribution, pore orientation and connnectivity [35]. In general, geotextiles are discontinuous materials and are being produced from macroscopic sub-elements, namely fibres or filaments consisting of void spaces or pores with finite porosities. Porosity is defined as the fraction of bulk volume of a material that is occupied by void space and can be more than 0.95 for bulky woven, knitted and air-laid nonwovens. It is an average physical parameter for any porous material, but the *real* structural characteristics of materials are not revealed. In other words, porosity is *incapable* of identifying the flow characteristics of any porous materials. However, pore volume is a key parameter in determining the capacity of liquid uptake. Pore connectivity is described in terms of geometric pathways that are dependent upon the orientation of pore in the structure.

Entrance and/or exit pore dimensions play a significant role in predicting the probability of soil particulates penetrating and being retained on the structure; pore dimensions at the entrance and exit may be different and thus described as the pore size distribution in geotextiles. It has been pointed out in the literature that pore size distribution is generally uni-and bi-modals in case of nonwoven and woven fabrics (i.e. manufactured from spun and multifilament yarns), respectively. Thus, appropriate distribution functions are required to study the broad range of pore dimensions in geotextile structures. For example, many researchers have established that the pore size distribution exhibit gamma distribution in a nonwoven structure [105–108]. It has also been established that the pore size in nonwovens is significantly affected by fibre orientation or alignment of fibres in the structure [105]. The nature of the measurement of pore size distribution, specifically for nonwoven geotextiles, is quite complex. Numerous techniques, namely dry sieving, hydrodynamic sieving, wet sieving, image analysis, bubble point and mercury intrusion porosimetry have been employed earlier to determine the pore size distribution of nonwovens but no single method has been accepted universally [109–115]. The principles of these methods for measuring the pore sizes are briefly summarised by Russell [35]. Nevertheless, Bhatia and Smith [115] have made a comparison between the various test methods for determining the pore size distributions of different types of geotextiles. They have found that the dry sieving methods yielded a larger pore size than the hydrodynamic and wet sieving methods. In case of slit film plain-woven geotextiles, the dry sieving and hydrodynamic sieving methods have given more accurate results of larger pore sizes in comparison to other methods. Moreover, bubble point and mercury intrusion porosimetry methods produced different results of pore size distribution. Nevertheless, the bubble point method has shown repeatability and was able to distinguish between thermal bonded nonwoven geotextiles of various thicknesses but there was little difference in the results of needlepunched nonwoven geotextiles. It is interesting to note that bubble point and mercury intrusion porosimetry methods are based on the principle of capillary flow, i.e. porous material allows the passage of the liquid when the applied pressure exceeds the capillary attraction of the liquid in the largest pore. The differences between the two methods existed in using the liquid which wets the geotextile in a bubble point test method, whereas mercury is a non-wetting liquid and will not enter into the pores of the material through capillary action. Therefore, it is anticipated that the contact angle between the liquid and the pore wall should be zero, specifically in the bubble point method. In general, liquids with high affinity for a solid will result in low contact angles and vice versa. Elton and Hayes [116] have studied the relationship between the contact angle and the pore size distribution of geotextile structures. They have noted that the change in pore size distribution as a function of contact angle is nonlinear in nature. It has also been demonstrated that lower pore size differed significantly when using different liquids however accounted for their surface tension and contact angle through dynamic contact angle (DCA) technique. Thus, it was concluded that some *unknown* property or phenomenon relevant to bubble point method has not been explored yet.

A comparison is made between the pore size of woven geotextiles obtained from the dry sieving method and image analysis technique and has been found to be a good agreement between the two test methods [117]. Furthermore, an empirical relationship between porosity and relative open area or cover factor has also been proposed and justified, specifically for woven geotextiles. The empirical relationship between the equivalent pore size of nonwoven fabrics (D), fibre diameter (d) and porosity (n) has also been established as shown in the following equation [118]:

$$D = 3.3dn. \tag{12}$$

An image analysis technique for measuring the pore size distribution has also been developed to analyse the 3D structure of a nonwoven geotextile [119]. The method has involved three stages, i.e. specimen preparation, image analysis and determination of pore size distribution. The larger pore size of nonwoven such as O_{95} (95% of the pores are smaller than this size) has shown a good comparison with the specification provided by manufacturer's reported apparent opening size. Image analysis method has also been used to relate the geotextile porosity as a function of normal stress [120]. In general, the needlepunched nonwoven geotextile porosity decreases with the applied normal stress and it was observed that at higher stress level of more than 1000 kPa, fibre to fibre contact increased resulting in decreased pore size, which further affected the filtration characteristics of geotextiles. Similar observations were made by Palmeira and Gardoni [121] for formulating the relationship between the porosity of a geotextile and the density of fibres at unconfined and confined conditions, as shown in Figure 9. Similarly, when tensile loading of 10% of tensile strength of a woven slit film was applied, the reduction in pore size decreased by 28% [122]. According to Fischer, Holtz, and Christopher [123], the application of small tensile loads caused the closing of smallest pores and subsequently largest pores also closed, which was found to be the probable reason for obtaining lower pore sizes when unequal biaxial loading was applied to a woven geotextile [122].

2.2.2. *Permeability characteristics of geotextiles*

The permeability characteristics of geotextiles requires the usage of fluid, such as air or water, and the volumetric rate of the fluid flow per unit cross-sectional area is measured and recorded against the defined differential pressures for obtaining the air or water permeability [35]. The air permeability is highly significant in geotextiles for the applications in which they are used as a gas collection layer in abandoned or uncontrolled old landfill sites [124].

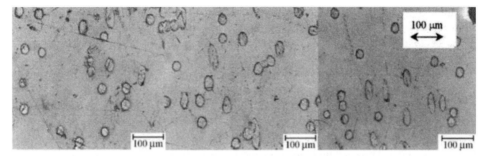

(a) Under 2kPa normal stress

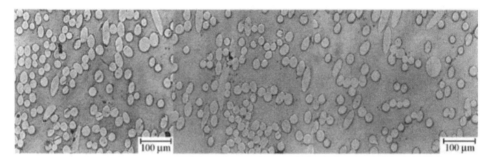

(b) Under 1000 kPa normal stress.

Figure 9. Cross sections of geotextiles under different normal stresses [121]. Reprinted from E.M. Palmeira and M.G. Gardoni, Geotext. Geomembranes 20 (2002) pp. 97–115, with permission of Elsevier.

The water permeability characteristics are of paramount importance primarily for filtration and drainage applications of geotextiles. The filtration and drainage functions require the passage of water in the in-plane and out-of-plane directions, respectively without the loss of soil particles. The water permeability is generally determined by the constant hydraulic pressure head and the falling hydraulic pressure head methods [35].

The flow of a fluid through a porous medium such as geotextile can be expressed by the coefficient of hydraulic conductivity, also known as the coefficient of permeability. It is generally normalised by the thickness of the geotextile for obtaining the permittivity and transmissivity in the cross-plane and in-plane directions, respectively as illustrated in the following Equations [125]:

$$k_n = \frac{q}{WL}\frac{t_g}{h}, \tag{13}$$

$$\psi = \frac{k_n}{t_g} = \frac{q}{WL}\frac{1}{h}, \tag{14}$$

$$k_h = \frac{q}{Wt_g}\frac{L}{h} \tag{15}$$

and

$$\theta = k_h t_g = \frac{q}{W}\frac{L}{h}, \tag{16}$$

where k_n is the cross-plane hydraulic conductivity, k_h is the in-plane hydraulic conductivity, q is the flow rate, h is the head loss in the flow direction, W is the geotextile width, L is the geotextile length, t_g is the geotextile thickness, ψ is the permittivity and θ is the transmissivity.

Permittivity and transmissivity are widely used in the engineering design but they do not allow a comparison of hydraulic conductivity of geotextiles with different thicknesses. Nevertheless, it is known that the thickness of geotextile decreases with an application of normal stresses. The thickness of geotextile also decreases with time due to creep under a constant compressive stress resulting in a decrease in transmissivity and permittivity [7,121,124–129]. Thus, a simple relationship has been formulated between transmissivity reduction and thickness reduction, as shown in the Equation (17) [126].

$$\frac{\theta_2}{\theta_1} = \left(\frac{t_2 - \frac{\mu}{\rho}}{t_1 - \frac{\mu}{\rho}}\right)^3, \tag{17}$$

where μ is the mass per unit area of the geotextile, ρ is the fibre density, θ_1 and θ_2 are the hydraulic transmissivities for corresponding thicknesses t_1 and t_2, respectively.

Several researchers have found that transmissivity and permittivity are highly dependent upon various parameters including raw material characteristics, process parameters, structural characteristics of geotextiles and level of normal stress [7,128,130–132]. According to Wei et al. [130], thinner nonwoven geotextiles yielded higher flow rates than the thicker nonwoven geotextiles as porosity apparently has a linear relationship with flow rates. However, the other important structural parameters of nonwovens including fibre diameter, proportion of fibres perpendicular to the fabric plane (specifically in a needlepunched nonwoven structure), fibre orientation and its distribution play a key role in determining the permeability characteristics of nonwoven geotextiles [133–135]. Similarly, in woven geotextiles, the structural parameters, namely yarn diameter, crimp, thread spacing and type of weave affect its permeability behaviour [131]. In general, woven geotextiles have poor permeability characteristics in comparison to the corresponding nonwoven geotextiles. According to Ling and Tatsuoka [125], the composite consisting of woven and nonwoven structures performed equally well, specifically in terms of in-plane hydraulic conductivity, as that of a nonwoven geotextile of similar density, although they have also reported that the coefficients of in-plane permeability of woven fabrics are similar to those of needlepunched nonwoven structures at low stress levels. Moreover, on increasing the normal stress levels, the coefficient of in-plane permeability of woven fabric is found to be lower than the needlepunched nonwoven geotextile [136]. Koerner et al. [127] have studied the air and water transmissivities of needlepunched nonwoven geotextiles and it was found that the air permeability of the geotextile was several order magnitude higher than that of water permeability because the viscosity and density of air were much lower than that of water. Furthermore, they found that the applied normal stress had no significant influence on the air transmissivity of geotextiles, which was quite contradictory to the in-plane water permeability tests. However, the gas transmissivity of needlepunched nonwoven geotextiles was investigated by highlighting the influence of moisture content and stresses [124]. It was demonstrated that the gas transmissivity decreased with the increase in moisture content and stresses. Thus, it can be summarised from the above discussion that the material properties, air-flow characteristics and stress levels can significantly change the behaviour of in-plane air permeability characteristics of geotextiles.

The average permeability of a geotextile does not provide information about its structural characteristics. In other words, flow in preferred directions due to fabric anisotropies or permeability variations due to structural heterogeneities are extremely important in hydraulic characteristics of geotextiles. Rebenfeld and his colleagues published a series of papers experimentally quantifying the directional permeabilities of nonwoven geotextiles [137–139]. It was pointed out that homogenous fabrics yielded circular or elliptical flow fronts, whereas heterogeneous fibrous networks deviated from the elliptical shape due to spatial variations in areal densities, irregularities in fibre orientation, variations in fibre compositions and variations in degree of needlepunching or inter-fibre bonding. Surprisingly, it was found that the viscosity of the fluid did not affect the directional permeabilities of nonwoven geotextiles [137]. Similarly, Mao and Russell developed a series of models of directional permeabilities of nonwoven geotextiles [133–135]. Moreover, it has been noticed that permeability distribution of heterogenous porous media, such as a nonwoven, is generally skewed. For instance, log normal distribution is found to be more suitable for its permeability distribution [140]. Nevertheless, the velocity of flow disturbed by the geotextile is extensively studied by means of a laser Doppler velocimetry technique [141,142]. It was observed that needlepunched structures had yielded smoother flows than the heat-bonded nonwoven geotextiles. Furthermore, in damaged geotextiles, at a given distance from the fabric, the flow velocity increases with the Reynolds number and the needlepunched nonwoven structure can absorb local defects, namely rips more easily than other types of geotextiles.

The cross-plane and in-plane hydraulic conductivities of geotextiles have also been investigated under operational conditions, particularly for reinforced soil structures [125,128,130]. Generally, it has been found that as the cake of soil particles tend to build up, the permeability decreases drastically with an increase in filter cake thickness. This is partly attributed to the soil penetration and retention in the geotextile resulting in closing off the flow channels or pores. According to Ling and Tatsuoka [125], the hydraulic conductivity of the nonwoven geotextiles was found to be the lowest in the soil-confinement test than in the block- and membrane-confinement tests, because it was apparent that the degree of the interfacial gap between the geotextile surface and the soil particles was much closer than that of other confining materials, namely rigid block and flexible membrane. Thus, it was established that the soil-confinement test is required for determining the in-plane hydraulic conductivity of a geotextile under operational conditions.

2.3. Chemical properties

When polymeric fibrous materials in the form of geotextiles are exposed to hostile environments, they often undergo structural changes that can be both morphological and chemical in nature, as the fibrous materials have a much higher surface area and as a result these changes may occur more rapidly than in the bulk polymers. The material vulnerability to an environment may be reflected in changes in crystalline or amorphous (non-crystalline) regions of the fibrous material and often these changes are accompanied by chemical modification through either hydrolytic or oxidative reactions. These are considered to be the degradation reactions and are catalysed by light and traces of metals present in polymers. Crystalline regions are more resistant to attack than the non-crystalline regions. Random chain scission of polymer molecules results in the deterioration of the material's mechanical properties. The chemistry of a polymer, repeat unit and functional group determines its sensitivity to the chemical degrading agencies as indicated in Table 5 [143]. For example, there are significant differences in the stabilities of aliphatic and aromatic chain

Table 5.   Degrading agencies and sensitive intrinsic polymer structure features [143].

Degradation type	Sensitive polymer chemical structure
Thermal degradation	Aliphatic groups and single bonds (nylon 6,6) Reactive side groups ($-CN$ groups in PAN)
Thermal oxidation	Polymer chain $-CH_2-$ (in melt processed synthetics)
Photo degradation	Presence of UV absorbing groups (aromatic ring structures in polyester and aramid)
Photo oxidation	Relates to presence of thermal degradation-generated chromophores in polymer chains
Chemical oxidation	Oxidisable hydrogen in aliphatic groups ($-CH_2-$, and $-CH_3$) Functional side groups ($-OH$, $-CN$)

fibre-forming polymers, presence of functional groups in the chains of polycondensation polymers (polyesters and polyamide) and the presence of reactive side groups in polymers such as polyolefins. However, the presence of impurities in the polymer can also lead to a change in the chemical stability of the material. For example, the antioxidants are used to reduce the effect of oxygen and other oxidising species during melt and heat processing of thermoplastic polymers such as polyester and polypropylene.

Geotextiles encounter a wide variety of environments in their applications and the influence of the chemistry of the environment is an important consideration when engineering the geotextile for different applications such as municipal and hazardous waste landfills, geotextile reinforcement of slopes and subgrades and earth retention systems. It is therefore important to understand the influence of these conditions on the behaviour of the selected geotextile. Both the natural and synthetic fibres are used in the production of geotextiles, as has been discussed earlier. However, polyester- and polypropylene-based geotextiles are the most important and commonly used materials for civil engineering applications. Consequently, these two types of geotextiles have been extensively studied from the long-term stability point of view. For example, in landfill applications, polypropylene nonwoven geotextiles are installed above the geomembranes for protection and drainage [144]. These geotextiles are exposed to chemicals in the form of acidic or alkaline solutions, especially the leachate solutions, until the reclamation of waste is completed [145]. In many countries most of the waste in municipal landfills is wet food and liquid waste that are either acidic or alkaline during the landfill periods. Therefore, the chemical resistance of geotextiles and other geosynthetics to the leachate solutions from different waste landfills has been assessed by many researchers [146–148]. Mathur, Netravali, and O'Rourke [149] have investigated the influence of pH and saline environments on various properties of geotextiles. These workers have also carried out accelerated aging experiments on polyester and polypropylene geotextiles at ambient and elevated temperatures. These tests have been 95°C for six months in saline (pH 8), strong alkaline (pH 10) and acidic (pH 3) media. The changes in the properties were determined using a variety of techniques such as tensile testing, differential scanning calorimetry (DSC), thermogravimetric analysis (TGA) and intrinsic-viscosity (IV) measurements. Results of these investigations show that polyester undergoes hydrolytic degradation under both acidic and alkaline conditions at and above the glass transition temperature, with the degradation being more severe under alkaline conditions. The polypropylene geotextile was found to be relatively inert to the pH conditions and showed no changes in strength. These workers also reported an increase in crystallinity in the initial period of ageing for both polyester and polypropylene. It was proposed that the increase in polyester crystallinity could be due to the preferential attack on the amorphous polyester regions and in the case of polypropylene the observed increase might be due to nucleation growth and lamellar thickening.

Polyolefin geotextiles are chemically stable products and are not much affected by the chemistry of the environment in which they are used. However, polypropylene is susceptible to oxidation during exposure to UV light, resulting in the deterioration of strength and increased brittleness of the geotextile. The effect is enhanced at elevated temperatures. The main mechanism of oxidative degradation is free radical and is initiated by thermal or photolytic cleavage of the polymer bonds. The thermal oxidation of polypropylene geotextiles has been investigated using three different methods [150]. Acceleration of the oxidation kinetics was obtained through increased temperature, increased oxygen pressure and the immersion of samples in aqueous media. The results of the study showed that high temperature ageing tests were less representative than tests at more moderate temperatures in which ageing was accelerated by high oxygen pressure and immersion in an aqueous medium. In a recent study, Palmeira, Remigio, Ramos, and Bernardes [151] have also carried out work on the oxidative stability of polypropylene and polyethylene geotextiles. In order to investigate whether the testing at high oxygen pressure and relatively low temperature could be an alternative method, polypropylene tape yarns for geotextiles were aged at enhanced oxygen pressures. The results showed that testing at high oxygen pressures could substantially increase the rate of ageing of geotextiles. Polyester-based geotextiles have good resistance to chemicals including organic acids, organic solvents, petroleum and bitumen. Polyester is susceptible to alkaline hydrolysis, but in soil environment, the alkaline conditions rarely exceed pH 10. However, in the presence of calcium ions, the rate of alkaline hydrolysis shows a tremendous increase. Therefore, in the hard water environment the strength of the polyester-based geotextile is expected to decrease rapidly with time. Behaviour of various types of geotextiles under soil burial conditions is shown in Table 6. In general, polyester fibres have moderate resistance to UV radiation, which is generally similar to polyamide fibres. Depending on the type of delustering agent, dyestuff and other additives incorporated into the polymer, the sensitivity of polyester can either increase or decrease. Alkaline water is known to affect the properties of polyester-based materials and this aspect has been investigated by Halse, Koerner, and Lord [152] in a study on the effect of high alkalinity levels on various types of polyester geotextiles. The workers prepared solutions of pH 10 and pH 12 (with pH 7 as control) and various geotextiles were incubated up to 120 days. Flow and strength tests were conducted on all test specimens on a weekly basis for the entire duration. The results showed that the alkalinity increased the flow time for a given volume of solution to pass through the geotextile and the effect was more pronounced for pH 12 than for pH 10 solutions. Formation of a precipitate was observed, which rested either on or within the fabric structure causing the major portion of the increased flow time. Furthermore, a reduction in the tensile strength of the geotextile was observed. It was also noted that some polyester fabrics showed measurable strength losses, while others showed no change. Nevertheless, polyester and polypropylene geotextiles have excellent chemical and physical properties and these materials can be used in almost any environment where long-term durability features are important criteria.

Table 7 summarises the essential physical, mechanical and chemical properties required for desired functions of geotextiles [4].

2.4. Process–structure–property relationships

It has been pointed out in Section 1.5 that more than 75% of the market utilises geotextiles made from various nonwoven fabrics, chiefly staple-fibre needlepunched and

Table 6.   Behaviour of geotextiles under soil burial conditions.

| Geotextile | | Duration | | |
Fibre type	Fabric structure	(years)	Soil type/location	Effect
Polyester	Woven	2–7	pH 4.7–11.5	5–15% strength loss
	Nonwoven	15	pH 9–10	12–40% strength loss
	Woven straps	17		Insignificant
	Woven	1	Peat	Up to 30% strength loss
	Nonwoven	7	Organic soil	Negligible
	Woven	3.3	PFA/water pH 8.7	3–6% strength loss
			PH 11.9	4–7% strength loss
	Nonwoven	1–6	Drainage	0–26% strength loss
Polypropylene	Nonwoven	4–6	pH 4.7–11.5	Up to 30% strength loss
	Woven	7	Water/bank interface	15% strength loss
	Woven	1	Peat	Up to 30% strength loss
	Woven	5–6	Permanent edge	20–30% strength/extension loss
	Nonwoven	7	Organic soil	Negligible
	Nonwoven	10	Railway/separation	20% strength loss
			Erosion protection	8% strength loss
Polyamide	Woven	7	Water/bank interface	30% strength loss
	Woven	1	Peat	Up to 30%
	Woven	10	Sea water	About 20% strength loss
	Woven	15	Bank mattress	35–60% strength loss – upper surface
				3–30% strength loss – lower surface

continuous-filament spunbonded nonwovens. These nonwoven structures are required to be engineered for the desired functions or applications. This can only be possible by using appropriate raw material, keeping the *right* process conditions and other related parameters. Thus, the effect of important process conditions/parameters of spunbonding and needlepunching on various properties of nonwoven geotextiles is discussed.

The process parameters can be classified in two categories, i.e. online and offline. Generally, offline parameters are kept constant and can only be changed when the production line is not in operation. In a spunbonding process, the main important online parameters include polymer throughput, polymer/die temperature, quench air rate and temperature, take-up speed and bonding conditions [153–158]. It has been reported that polymer throughput rate and die/polymer temperature control the diameter and texture of the filament. Bonding temperature and pressure play a significant role in determining the tensile properties of spunbonded structures. Quenching conditions are mainly responsible for cooling of the filaments and hence the development of microstructure, whereas take-up speed controls the draw-ratio and filament orientation or deposition on the conveyor belt. Several researchers have studied the process of laying down of filaments onto a moving belt [159–162]. It was demonstrated that filament properties such as linear density, bending rigidity, torsional rigidity etc. in addition to process parameters, namely height of the feed point and ratio of feed speed to the conveyor belt speed controlled the filament orientation in the spunbonding process. The filament orientation dictates the physical, mechanical and hydraulic properties. Moreover, isotropic characteristics of nonwoven structure are highly dependent upon

Table 7. Functional requirements for geotextiles [4]. Reprinted from A.R. Horrocks and S.C. Anand (eds.), *Handbook of Technical Textiles*, Woodhead, Cambridge, UK, 2000, with permission of Woodhead Publishing Limited, UK.

Geotextile function	Tensile strength	Elongation	Chemical resistance	Biodegradability	Flexibility	Friction property	Interlock	Tear resistance	Penetration	Puncture resistance
Reinforcement	iii	iii	ii–iii	iii	i	iii	iii	i	i	i
Filtration	i–ii	i–ii	iii	iii	i–ii	i–ii	iii	iii	ii	ii
Separation	ii	iii	iii	iii	ii	i	ii	iii	iii	ii
Drainage	na	i–ii	iii	iii	i–ii	na	ii	ii–iii	iii	iii
Erosion control	ii	ii–iii	i	iii	iii	ii	i	ii	ii	i–ii

Geotextile function	Creep	Permeability	Resistance of flow	Property of soil	Water	Burial	UV light	Climate	Quality assurance & control	Costs
Reinforcement	iii	na–i	i		iii	iii	ii	na	iii	iii
Filtration	na	ii–iii	i	ii	iii	iii	na	iii	iii	iii
Separation	na	ii–iii	i	na	iii	iii	na	i	ii	iii
Drainage	na	iii	i	na	iii	iii	na	iii	iii	iii
Erosion control	na	ii	iii	na	iii	na	iii	iii	i	iii

Notes: iii = highly important, ii = important, i = moderately important, na = not applicable.

the filament orientation in a spunbonded structure. The offline spunbonding parameters include screw parameters, spinneret hole size, spinneret to collector distance and bonding system.

In a needlepunching process, there are many variables that can affect the properties of geotextiles. The list includes needle arrangement in the needleboard, effective length of the needleboard, direction of needling (top/bottom or both), diameter of the holes in stripper and bed plates, distance between stripper and bed plates at the entrance and exit, barb shape, number of barbs on a needle, barb spacing, barb angle, kick-up, depth of needle penetration and punch density [35,163–165]. Some of these parameters (specifically barb parameters) are kept constant, although they can significantly influence the performance characteristics of geotextiles. For example, the permeability of geotextiles can be enhanced by selecting a coarser gauge blade and using larger barbs, a higher kick-up and an open barb [35]. Similarly, the geotextile thickness can be increased by choosing smaller and fewer barbs per needle, decreasing the barb angle and increasing the barb spacing. Nevertheless, the most important process parameters influencing the geotextile properties are the depth of needle penetration and punch density. The depth of needle penetration is defined as the distance the needle point passes below the top surface of the bed plate [167]. It determines the number of barbs penetrating the web and thus affects the degree of fibre entanglement and bonding in a needlepunched structure. The punch density (P_d) is defined as the number of needle penetrations per unit area. It is a function of fabric output speed (O), number of needles per unit width (n) and stroke frequency (S) as shown in Equations (18) and (19). Generally, the number of needles is kept constant and punch density is mainly dependent on the stroke frequency and output speed.

$$P_d = \frac{n}{A} \tag{18}$$

and

$$A = \frac{O}{S}, \tag{19}$$

where A is the advance per stroke.

Hearle and Sultan [163], and Hearle et al. [164] have reported that the amount of needling and the depth of needle penetration have a significant effect on the physical and mechanical properties of needlepunched nonwoven structures. Increasing the depth of needle penetration resulted in a decrease in the fabric area density and breaking extension, initial increase in the modulus and tenacity up to a certain level, which finally decrease at higher needle penetration [164]. The modulus and tenacity increases with an increase in the number of fibres in the thickness direction and subsequently there is fibre damage due to the increase in the transportation distance of fibres [166]. Similar effects can be observed for the effect of punch density on various properties of needlepunched nonwoven structures [164]. The effect of the punch density and depth of needle penetration have also been studied on physical properties of geotextiles made from polyester and polypropylene fibres [167]. However, these studies did not report simultaneous influence of more than one parameter.

Rawal and Anandjiwala [168], Rawal, Majumdar, Anand, and Shah [169] and Rawal et al. [58] have published a series of papers reporting the combined effect of process parameters including stroke frequency, depth of needle penetration and punch density on various properties of needlepunched nonwoven geotextiles. It was observed that the effect of the depth of needle penetration was more significant than punch density on fabric thickness whereas higher punch densities were relatively more effective in consolidating the higher web area density leading to a sharp increase in fabric area density [169]. Similarly, puncture resistance of needlepunched nonwoven geotextile was found to be more dependent upon punch density when the combined effect of punch density and depth of needle penetration were analysed. Furthermore, isotropic, needlepunched nonwoven structures could be produced at higher punch densities for a given range of web area densities. The effect of stroke frequency and the depth of needle penetration on fibre orientation were also studied and it was noted that an increase in the depth of needle penetration caused an increase in the number of fibres in the machine direction, whereas most of the fibres were initially orientated in the cross-machine direction [168]. However, an increase in the stroke frequency reduced the number of fibres in the machine direction. Moreover, an increase in the stroke frequency or depth of needle penetration reduced the pore size of geotextiles. It is anticipated that fibre damage at higher depth of needle penetration or stroke frequency can cause disintegration into small constituents resulting in blocking-off the pores in the structure. The permittivity and permeability coefficients of staple-fibre needlepunched nonwoven geotextiles were found to be reduced with increase in the depth of needle penetration and punch density [129]. In the case of spunbonded needlepunched nonwovens, the permittivity initially increases and then attains a constant value with increasing depth of needle penetration and punch density. However, the transmissivity of needlepunched nonwoven geotextiles decreases with an increase in the stroke frequency [168].

2.5. Test methods of geotextiles

Some of the important test methods for measuring the geotextile properties are given in Table 8 [35].

3. Performance of geotextiles

The performance of geotextiles can be assessed through the design criteria and durability of geotextiles as discussed below.

3.1. Design of geotextiles

The design of geotextile is of paramount importance for any civil engineering application. In general, the design test methods are superior to control or index tests as the former tests are performed using the boundary conditions that simulate the field conditions, whereas the control tests are performed to evaluate the quality of the product and index tests are used for comparing different types of geotextiles. The results of index tests can be used by an experienced designer using an appropriate factor of safety in order to replace design tests as the latter are expensive methods. However, the index test can be responsible for failure of geotextile due to the lack of evaluation of field performance [8]. These index and design tests along with designers' experience and cost information can be used for determining the performance criteria for various geotextiles. For instance, it has been demonstrated that the selection of the geotextile for drainage applications is a part of the design process and the ranking of geotextiles can be given in the following categories, i.e. filtration and separation efficiency, flow capacity, stress–strain behaviour, chemical compatibility, cost, durability and survivability [8]. The decision is, however, affected by the prejudice of the engineer or designer performing the analysis.

Table 8. Test methods to measure some important properties of geotextiles [35]. Reprinted from S.J. Russell (ed.), *Handbook of Nonwovens*, Woodhead Publishing, Boca Raton, FL, 2007, with permission of Woodhead Publishing Limited, UK.

Geotextile	North American (ASTM, AATCC and MIL)	EDANA and INDA (WSP, ITS and ERT)			Europe (ISO, BS and EN)
		ITS	WSP	ERT	
Geotextiles – vocabulary					ISO 10318:1990
Guidelines on durability					ISO/TR 13434:1998
Sampling	ASTM D4271-01	ITS 180.1			
Mass per unit area	ASTM D5261-92				EN 965:1995
Thickness					EN 964-1:1995
Breaking (grab strength)	ASTM D5034-95	ITS 110.1			
Trapezoid tear	ASTM D4533-04	ITS 180.3			
Puncture strength	ASTM D4833-00el	ITS 180.4			EN ISO 12236:1996
Dynamic perforation (cone drop test)					EN ISO 918:1995
Bursting strength	ASTM D3786-01	ITS 30.1			
Pore size	ASTM D4451-04 (D6767-02)	ITS 180.6			EN ISO 12956:1998
Permittivity	ASTM D4491-99a (D5493-93)	ITS 180.7			EN ISO 11058:1998
In-plane transmissivity	ASTM D6574-00				EN ISO 12958:1998
Thermoplastic fabrics in roofing/waterproofing	ASTM D4830-98	ITS 180.8			
Wide-width tensile test	ASTM D4595-86	ITS 180.9			EN ISO 10319:1996
Abrasion damage simulation (sliding block test)	ASTM D4886-88				ISO 13427:1998
Determination of the resistance due to weathering (UV)					EN 12224:2000
Determination of the resistance to acid and alkali liquids					EN 12960:2001
Determination of resistance to oxidation					EN 13438:2000
Resistance to microbiological attack by soil burial test					EN 12225:2000
Procedure for simulating damage during installation					EN 10722:1998
Evaluation following durability testing					EN 12226:1996

According to Leflaive [170], the consultants will not include geotextiles in their designs unless they have reliable technical data and design methodology for these materials. Similarly, Koerner [171] has suggested that the selection of woven or nonwoven geotextile should be realised by the "design by function" route so that the properties of geotextiles are specified. Hence, some of the important design methods for different functions of geotextiles are discussed below.

Geotextile filters have successfully replaced graded filters over the last few years due to their comparable performance, improved economy, consistent properties and ease of replacement. These characteristics of geotextile filters are mainly dependent upon their design for which these can be used for drainage and filtration applications. The list of applications include pavement edge drains, dewatering trenches, armoured slopes and shorelines, prefabricated drainage panels, leachate collections systems and landfill caps [172]. The design of geotextiles as a filter is similar to that of graded granular filters, i.e. it consists of soil particles with voids, whereas geotextiles contain filaments of fibres along with pores. As discussed earlier, the pore shapes and sizes are quite complex in geotextiles; however, pore size can be measured experimentally and when compared with soil particle size, the filtration characteristics of geotextiles can be determined by obtaining the retention capacity of geotextile for soil particles while maintaining the required flow. This has led to the following three simple filtration principles for geotextiles to be an effective filter [173]. Firstly, the soil particle will not pass through the filter if the size of the largest pore in the geotextile is smaller than the larger particle of the soil. Secondly, the geotextile would not get clogged if the majority of the openings in the geotextile were sufficiently larger than the smaller particles of soil. Finally, a large number of openings are the pre-requisite to maintain the proper flow through the geotextile. Furthermore, three distinct mechanisms of filtration have been recognised in view of the geotextile being used in drainage systems and other earth-related structures [174]. These filtration mechanisms are: geotextile in contact with soil, particles in suspension, and soil particle migration under cyclic loading. When the geotextile is in contact with soil particles, it must act as a barrier and/or in combination with a natural filter using *self-filtration* or *vault network formation*. In the *self-filtration* mechanism, the coarser soil particles form a layer at the geotextile interface and thus the soil particle migration is stopped. However, in the case of the *vault network* type of filtration, the particles arrange themselves in the form of vaults at the filter interface due to the electrical and adsorption forces existing between lubricant and antistatic agent on the fibres/soil particles and between the particles. *Particles in suspension* type of filter consists of soil particles that are carried by water and upon approaching the filter forms an impervious layer, which considerably reduces the flow rate in a given time and the filter is thus required to be replaced. In the *cyclic loading* type, the smaller soil particles tend to migrate towards the geotextile due to the hydraulic pressure being generated at the contact points and the particle migration at the filter interface takes place with ease. These filtration principles and mechanisms form the basis of design criteria of a geotextile. The design criteria are discussed below.

- *Retention criteria*: The retention criteria for geotextiles are similar to that of soil filter criteria [173]. The geotextile is designed such that larger particles are retained to form a soil "bridge" resulting in the development of a stable soil structure, capable of preventing further migration. This forms the basis for the *concept of positive retention* [175]. On the other hand, complete retention of soil can only take place if the geotextile pores can retain the smallest particles and both concepts can be analysed by determining the relationship between the effective pore size of the geotextile and soil particle size [176]. Generally, the geotextile pore size is related to the base soil particle size as shown in the following equation [177]:

$$O_n < \lambda d_x, \tag{20}$$

where O_n is the geotextile pore opening size, λ is the constant dependent upon the surrounding conditions and d_x is the base soil particle size.

Various reviews and summaries of notable retention criteria are presented by several researchers [120,173]. These critical reviews for determining the "right" retention criteria can confuse engineers and designers due to the associated problems. Some of these problems and related solutions have been discussed by Christopher and Fischer [173]. Firstly, there is a disagreement on the effective pore size in the geotextile for retention, i.e. O_{95}, O_{90}, O_{50} or O_{15}. A solution was proposed that a better understanding of test method or procedure should be reviewed before applying to check its compatibility with the specific project conditions. Secondly, the determination of the effective opening size of geotextile is questionable and some of the related problems and solutions have been discussed in Section 2.2.1. Finally, the "*idealised*" filter bridge can only occur if there is intimate contact with the soil, but it can be highly affected by external factors such as reversing flows, dynamic conditions or movement of geotextile between riprap etc. Moreover, the filter bridge assumption is invalid for soils which are internally unstable. A probabilistic model to predict the retention capacity of a nonwoven geotextiles has also been developed [179]. The influence of pore size distribution, porosity, thickness and compressibility on the retention capacity of a nonwoven geotextile filter was analysed. One of the limitations of this model is that the pore size distribution of the geotextile is required to be known. Alternatively, a mathematical model of mass of soil particles passing through per unit area of geotextile has revealed that the filtration performance of geotextile is dependent upon geotextile structure, soil characteristics and filtration parameters [180]. The model, however, did not account for continuous reduction in pore size during the process of filtration. Similarly, a simple relationship between the yarn diameter of the woven geotextile and the diameter of soil particle has been formulated for various types of weaves [181]. The retention capacity of geotextile not only depends upon its structural characteristics but is also affected by soil properties (type of soil, coefficient of uniformity, coefficient of curvature, density index etc.) and flow conditions [117,121,176,182].

- *Permeability criteria*: One of the most important filtration principles is that the geotextile must have high permeability so that there should not be a build up of excess pore water pressure [173]. In general, geotextiles need to be more permeable than the soil it is retaining based on the assumption that the flow should not be hindered at the soil/geotextile interface [183,184]. In the past, it has been recommended that it is safe to keep the initial permeability high so as to ensure that the geotextile has the adequate permeability over its lifetime [185]. However, Giroud [178] has suggested that the geotextile permeability could only be 10% of the soil permeability assuming that the soil was thicker than the geotextile. A geotextile with 10% permeability to that of the soil is still expected to have a greater flow capacity than the soil, which is quite contentious. Thus, the main problem with the permeability criteria is to judge the permeability of the geotextile in relation with that of the soil. Christopher and Fischer [173] have suggested that an assessment should be made in terms of clogging potential and the *factor of safety* can be increased accordingly.

- *Clogging criteria*: Clogging is caused by the penetration of fine particles in the geotextile resulting in the blocking of pores or caking up the upstream side of the geotextile and a progressive increase of the water head loss in the geotextile [173,186]. In general, clogging and permeability criteria are closely related to each other, and in the past it was observed that retention and permeability criteria did not necessarily provide for a complete filter design as the system could still fail by clogging [185]. Luettich et al. [172] have suggested that the geotextile with many openings should be preferred in comparison to the geotextile with fewer openings as some of the openings can be blocked by soil particles and it is anticipated that the remaining openings are still available for maintaining the adequate permeability characteristics. Alternatively, a lower limit for a filter criteria can be based on positively lose soil particles that ensures adequate permeability and resistance to clogging, which forms the basis for the *concept of positive wash through* [175]. A relationship between particle size to both the diametric and volumetric pore size distributions of a geotextile is required to be determined for preventing the clogging. Several researchers have formulated the relationships between the clogging, porosity and pore size distribution and some of these have been reviewed and summarised by Christopher and Fischer [173]. However, none of them addresses clogging by simple criterion, as the basic criterion for clogging commonly involves the performance of filtration tests on specified soils. Nevertheless, for less critical applications, some designers have suggested to control the effective porosity and/or the smaller pore sizes of the geotextile. Luettich et al. [172] have recommended that nonwoven geotextiles should not have less than 30% porosity, whereas in woven geotextiles the minimum percent open area (POA) should be 4% in order to minimise the risk of clogging. Various tests can be performed to evaluate the performance of filter behaviour with a given soil and have been summarised by Sansone and Koerner [187], as shown in Table 9. Alternatively, an empirical model has been developed to simulate the accumulation of fine particles in a geotextile filter and a relationship between the water head loss and injected particles has been established [186]. However, some of the parameters in the model have been arbitrarily assigned in order to match the theoretical results with the experimental values. The major issue with these tests and models is that they only provide results for predefined soil and geotextile systems. Moreover, filtration tests are quite complex and expensive, consequently limited tests are performed, which may not simulate the *realistic* field conditions. Christopher and Fischer [173] have suggested that the characteristics, namely pore size distribution, porosity and filtration length are required to be defined and a universal geotextile filter criteria should be developed accordingly.
- *Survivability and durability criteria*: None of the above filter design criteria will be effective if the geotextile is damaged under construction or during installation [172,173]. Thus, the geotextile should exhibit minimum index strength properties corresponding to the severity of installation. Luettich et al. [172] have provided the guidelines in selecting the required mechanical properties of geotextiles depending upon the degree of installation. Similarly, the geotextile filters are exposed to sunlight and in such cases additives such as carbon black or titanium dioxide are recommended to provide the necessary resistance due to UV light. Some of the effects of UV radiation and additives are discussed in Sections 2.3 and 3.2.

In most of the filtration criteria, a characteristic pore size needs to be used. This pore size can be in the form of apparent opening size (AOS), equivalent opening size (EOS),

Table 9. Laboratory test methods for assessment of soil clogging and retention of geotextile filters [187]. Reprinted from L.J. Sansone and R.M. Koerner, Geotext. Geomembranes 11 (1992) pp. 371–393, with permission of Elsevier.

S. No.	Method	Principle	Advantages	Disadvantages
1.	Long-term flow (LTF)	Flow downward through soil and geotextile at constant head to measure flow rate over time	Straightforward setup Simple measurements Direct conclusions	Long time for equilibrium Biological growth in system Uncontrolled stress conditions
2.	Gradient ratio (GR)	Flow downward through soil and geotextile at constant head to measure hydraulic gradients	Crops of engineers initiated ASTM Standard D5105-90 Rapid results	Air in piezometer tubes Instability for fine grained soils Uncontrolled stress conditions
3.	Hydraulic conductivity ratio (HCR)	Triaxial permeameter without, then with, a geotextile to calculate a permeability ratio	Uncontrolled stress conditions Back pressure saturation possible	Requires triaxial soil Requires time to establish flow equilibrium Relatively new technique
4.	Dynamic filtration (DF)	Slurry of soil is added to a saturated geotextile system under dynamic conditions to note flow rate behaviour	Integrated into design method Simulates pulsed loads Soil can cover geotextile or be in-isolation Rapid results	Not representative of many situations Test setup quite complex Extremely challenging test conditions Very new technique
5.	Fine fraction filtration (F³)	Slurry of the fine portion of soil is directed at a saturated geotextile system	Fines are examined directly Soil piping readily assessed Rapid results	Not representative of entire soil High hydraulic conditions Vary new technique

filtration opening size (FOS) and effective opening size (D_w) [174]. It is important that porometry of a geotextile should be related to the hydraulic conductivity. Various institutes and committees such as the French Committee on Geotextiles and Geomembranes (CFG) criteria, Franzius Insitute of Hannover (FIH) criteria, Giroud criteria, Federal Highway Administration criteria and Ecole Polytechnique of Montreal (EPM) criteria, have formulated different filtration criteria and these criteria were critically discussed by Rollin and Lombard [174]. These criteria are useful if a geotextile performs well according to the intended application. Faure, Farkouh, Delmas, and Nancey [188] have analysed the behaviour of geotextile filter after 21 years. The geotextile was used as a filter in Valcros dam in France, and interestingly the performance of geotextile has not changed significantly since its installation. The outflow water was free from soil particles and the permeability of geotextile was greater than the soil to be protected. It was also found that 40 to 80% permeability of geotextile is sufficient to maintain its filter function.

The drainage performance can only be improved when the major functions such as filtration, drainage and separation interact with each other [189]. Separation is significantly important in determining the performance characteristics of geotextiles. As mentioned earlier, a geotextile as a separator precludes the mixing of coarse grains, a constituent of base course, with subgrade soil comprising sand, silt or clay and also prevents the intrusion of fines from the subgrade into the base layer under certain loading conditions. Hence, the opening or pore size of the geotextile should be such that it is able to prevent subgrade fines from moving up into the base layer under dynamic loads for both dry and wet conditions [190]. According to Narejo [190], geotextiles with opening size less than 85% of the size of soil would function adequately for separation. However, for fine silt and clay soils, the apparent opening size of a geotextile should be calculated as 0.5 times the 85% size of the soil. Furthermore, when the geotextile is loaded and pressed against the aggregate, the geotextile should be able to resist well against puncturing. A case study demonstrating the use of geotextile in a state highway with a poor pavement performance has been found to be effective in preserving the integrity of the system after five years of installation [191]. The subgrade soils at the test site have consolidated after the installation of geotextile in comparison to the sections without geotextiles. In this case, it was concluded that the long-term performance of geotextile separators might not be critical due to the increased subgrade strength and reduced compressibility due to consolidation. Similarly, the use of geotextiles in other engineering applications has become evident and one such application is reinforcement, i.e. the geotextile acting as a tension element being placed between the granular fill and the soft clay [192]. The design criteria for the geotextile to be used as reinforcement is quite complex, as the behaviours of such systems are dependent on the interaction of the granular fill, subgrade, geotextile inclusion and their associated strain compatibility requirements. Therefore, deformation of all components is difficult to be considered in a simple model that can be used in routine design calculations. Hence, simple design charts have been formulated based on semi-empirical procedures [64,193,194]. However, the inclusion of a geotextile can enhance the bearing capacity of a soil–geotextile system in the following three distinct ways, although in practice these function simultaneously [192]: (i) geotextile changes the failure mode, it tends to force a general rather than a local failure (see Figure 10(a)); (ii) geotextile tends to distribute a concentrated load over a larger area of subgrade, as shown in Figure 10(b) and (iii) geotextile provides a supplementary support due to membrane effect, the deformed geotextile provides an equivalent vertical support (see Figure 10(c)). In practice, the most important function is the membrane reinforcing and it is related to the load distribution by bulging, which is further dependent upon the stiffness of the geotextile [195]. More recently, Espinoza and

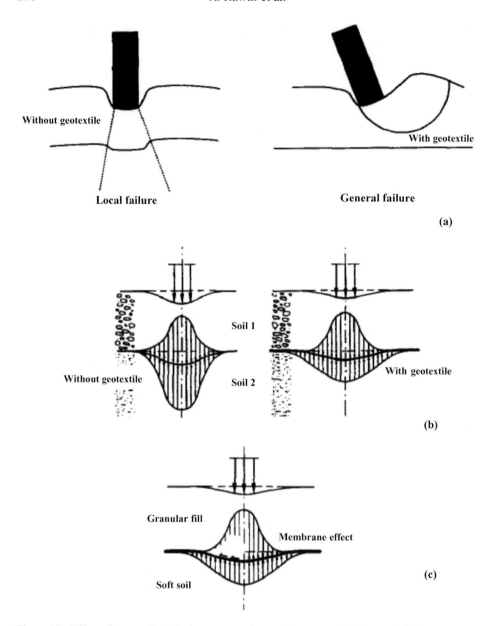

Figure 10.   Effect of geotextile inclusion on a two-layer soil system. (a) Change of failure mode. (b) Redistribution of applied surface load. (c) Membrane effect [192]. Reprinted from R.D. Espinoza, Geotext. Geomembranes 13 (1994) pp. 281–293, with permission of Elsevier.

Sabatini [196] have demonstrated that maximum tensile strength and modulus or strain can optimise the contribution of the reinforcement. For soft clays, it has been proposed that the maximum allowable strain in the fabric should not exceed 25% of the ultimate strain for permanent construction and 50% for temporary structures [197]. Furthermore, it has been published that woven geotextiles bond well with granular soil even for slope reinforced applications [198]. In general, the reinforcement in a slope can contribute to the stability in two ways, i.e. the shear resistance of the soil to resist the increased shear loading by the steep face is improved and it should act to hold the unreinforced soil mass in equilibrium without overstressing the underlying soils. The required forces for equilibrium depend upon slope geometry, soil properties, pore water pressures and applied surface loadings. In addition to the equilibrium concept, the required number of reinforcement layers can be added into the slope to provide the desired forces. In other words, the design criteria for reinforced slopes can be stated as the "provision of sufficient reinforcement layers, suitably distributed, so that the available force at every point exceeds the required force" [198]. The magnitude and distribution of required forces are dependent on material properties and safety factors, a concept adopted from *soil mechanics*.

In the past, different types of geotextile methods/criteria have been established by the manufacturers, national standards, educational and research institutions, and hence different approaches are required to be adopted which makes it difficult for an engineer to design and select a geotextile for the desired application. Furthermore, these approaches vary from country to country and according to Rigo, Mathieu, Smolders, and Alexandre [199], there are 300 different test methods for geotextiles available in the world. However, this limitation of designing a geotextile for a required application can be overcome by a simple computer program in the form of an *expert system*. Rigo and his colleagues have developed an expert system for the designing of geotextiles for various applications and the program is mainly divided into three main parts, i.e. selection of geotextile, user's personal requirements and database of geotextile characteristics [200,201]. Similarly, Mannsbart and Resl [202] have developed an expert system based on KnowledgePro (issued by Knowledge Garden Inc., USA) for various applications of geotextiles including unpaved roads, hydraulic construction, drainage systems, retaining walls and geomembrane protection. These expert systems can only provide guidelines to a designer or engineer but an in-depth understanding of textile and civil engineering is essential to solve innumerable construction and design problems.

3.2. Durability of geotextiles

The durability of a material can be defined as the ability of a material to remain intact and to carry out its prescribed functions effectively during the entire life of the project. In the case of geotextiles, the durability is determined mainly by the resistance exhibited by the constituent fibres to the environmental conditions. A geotextile should have sufficient strength and ability to resist when stretched, ruptured, punctured, as well as during cutting, compression, abrasion and silting. It should also have the required filtration characteristics, sufficient hydraulic resistance, low clogging and moisture absorption. In addition, the geotextile must be biologically and chemically stable and resistant to the effects of UV radiation. There are many degrading agencies to which geotextiles are exposed at the time of use. These materials are seldom exposed to a single degrading agency and the factors frequently act in combination, often in a complex manner and are little understood. Many civil engineering applications of geotextiles require designing of long-term performance and it should be noted that it is affected by handling, storage and installation of geotextiles [203]. Hence, the properties are chosen to ensure that the geotextile is suitable to maintain performance through transportation, storage, installation and loading and to perform its function during service. However, the geotextiles may become exposed to environmental factors at the time of installation and during the working life of the project and this aspect needs to be considered when choosing a geotextile to ensure it has sufficient physicochemical durability. The major environmental factors to be considered are temperature, exposure to UV radiation, pH, humidity and chemical conditions of the soil and the construction materials. Koerner [146] has reviewed the adverse effects of the long-term phenomena on the performance of geotextiles and related materials. They subdivided these into the effects of clogging, creep, chemicals, micro-organisms, light, burial and cumulative aspects.

As mentioned earlier, polyester and polypropylene fibres are widely used in the production of geotextiles due to their high biological and chemical stabilities [204]. Geotextiles made of these fibres lose only 10 to 20% of the initial strength due to natural ageing of the polymers in the first 3–5 years of use in complex natural climatic conditions. However, the disadvantage of polyester fibres is that exposure to calcium hydroxide (basic medium) significantly reduces their strength. This predetermines the unsuitability of these fibres for applications in geotextiles, where they come in contact with lime or concrete layers (for example, in the construction of tunnels). On the other hand, a polyester fibre geotextile is relatively insensitive to hot bitumen, since the melting point of the fibres is approximately 260°C. Similarly, the weathering stability of non-stabilised polypropylene is a major problem, especially if the geotextile is exposed to UV radiation at elevated temperatures. Thus, the non-stabilised polypropylene-based geotextiles should not be used in hot climates for long-term applications, especially if the geotextile is exposed or used near the surface. However, the use of stabilisers has solved this problem to a differing degree of success. The stabilisers have been developed to interfere with or modify the oxidative reaction of polypropylene that is initiated via its labile tertiary hydrogen bond in the repeat unit. The development of UV stabilisers is based on benzophenone and benzotriazole structures, which act as UV absorbers. More recently, hindered-amine light stabilisers (HALS) have also been developed. Carbon black is also an effective stabiliser and is extensively used to enhance the UV performance of polypropylene geotextiles, which may be exposed to UV radiation during their installation and lifetime. Furthermore, the durability of geosynthetics exposed to outdoor UV exposure has been investigated by various researchers [205–208]. Brand and Pang [205] placed 12 different geotextiles under Hong Kong outdoor exposure conditions and found that the half-life of the tested geotextiles ranged from three to more than nine months. McGown et al. [206] conducted a durability study of various types of geogrids and geotextiles in Kuwait region and observed very small changes in the mechanical properties of the tested PP and HDPE geogrids under the influence of UV radiation and heat cycling. However, the test results showed that the deterioration of exposed geotextiles can be significant, and the index test data do not closely reflect the long-term load-elongation behaviour of geotextiles. Cassady and Bright [207] have performed outdoor ageing experiments on polypropylene geotextiles and concluded that a minimum additive concentration is required to preserve the long-term mechanical properties of these geotextiles. The study concluded that carbon black, at 2.5% by weight, is the most effective means for retarding the deteriorating effects of UV light.

There is a growing use of geotextiles in marine environment, particularly for coastal defence applications. For example, geotextiles are placed below riprap along coastlines or harbour caissons to prevent soil erosion. Geotextiles may be immersed in the sea, placed in tidal zones or exposed to sunlight. In marine environment, geotextiles are exposed for long periods to the separate or combined action of seawater, UV light and temperatures ranging from arctic to tropical regions. Since the design lives of these structures are typically at least 25 years, the endurance and weathering properties of geotextiles in ocean environments are important for these applications. The effect of UV radiation and seawater on polyester and polyamide yarns has been studied by Horlker, Vevers, and Warwicker [209]. The researchers examined the influence of separate and combined action of seawater, air, oil and UV light on polyester and polyamide yarns through changes in their tensile properties and abrasion resistance characteristics. The results showed that breaking elongation and initial modulus values of both types of yarns were more significantly affected than their breaking load. The loss in abrasion resistance was more severe than the loss in tensile properties of both yarns types, and the effect of seawater on polyamide yarn abrasion resistance was more pronounced. Actinic degradation of polyester was enhanced by the presence of titanium dioxide in the yarn and the effect of UV light was not confined to the fibre surfaces only. In a more recent study, Hsieh, Wang, and Chiu [210] have also investigated the durability of polypropylene and polyester geotextiles in marine environment. They studied the effect of weathering conditions and reduction factors for two geotextiles in ocean environments. The study included outdoor UV exposure tests performed using a black polypropylene and a white polyester geotextile. The results indicated that UV exposure was the most significant factor in reducing the strength of geotextiles tested in an ocean environment. However, the effect of seawater immersion alone had the least weathering effect on the mechanical properties of the tested geotextiles. The study also showed that the effect on the mechanical properties of geotextile samples placed within the tidal zone was slightly higher than that for the immersion conditions.

Geotextiles have been extensively used in drainage and filtration of geotechnical and geoenvironmental projects. The marked growth of geotextile applications in environmental projects has also demanded research on these products, particularly in leachate collection systems of landfills. Landfill leachate is generally a high-strength wastewater characterised by extremes of pH, chemical oxygen demand (COD), biochemical oxygen demand (BOD), inorganic salts and toxicity [211]. In general, it is a complex material, which can cause physical, chemical and biological damage to filters by clogging, because it contains a mixture of various chemical compounds originating from different types of disposed materials and/or they may also be generated from biotic and abiotic processes in the system [212]. Many researchers have reported granular and geotextile clogging under laboratory/field conditions [128,151,186,213–218]. In the laboratory, most of the studies investigated the performance of geotextile filters in filtration devices where either real or synthetic leachate was used. These studies have demonstrated that biological clogging is a complex mechanism that can cause failure of a drainage system to fulfil its role in a landfill.

The leachate is generated because of liquid flowing through the soil, forming a solution of both dissolved and suspended materials. The composition of the leachate is highly dependent on the nature of the soil, type of waste and weather conditions because there are many parameters, which affect the solubility of organic and inorganic constituents present in the soil and waste. The range and level of the leachate is reported to be the highest in solid waste landfills [219]. The long-term leachate immersion tests were carried out in the laboratories of Geofabrics Limited, UK. Samples of polypropylene and polyester geotextiles were immersed in the leachate at an elevated temperature in a stainless steel tank. The samples were then tested at increasing time intervals for their puncture resistance in order to determine any change in the performance. It was noted that the strength of the materials increased initially by about 25% and then reduced and stabilised to 20% above the original performance of the geotextile [219]. The authors have concluded that the increase in the strength of the geotextile was probably due to three factors: the entrapment of suspended particles in the matrix of the geotextile, which increases the nodal friction between the fibres; the microbiological growth between the fibres, and stiffening of the fibres due to the increase in tank temperature. The increase in the strength was accompanied by a reduction in the breaking extension values.

There is a great ongoing discussion on the use of recycled polymers for the manufacture of geotextiles, because a significant amount of low-cost recycled fibre/polymer is available, which can be utilised for this purpose [220–223]. However, the long-term durability of the recycled material-based geotextiles must be established if these are to be employed in critical applications. The recycled fibre is often of mixed type or origin and therefore the geotextiles produced from this fibre may not have the same performance and consistency as the virgin material-based geotextiles. Davies and Horrocks [223] have extruded a series of polypropylene filaments containing recycled polyolefin materials in varying concentrations in the absence and presence of thermal stabilisers. Physical and chemical properties of the resulting polypropylene filaments were studied in order to determine the influence of recycled material on the properties of the resulting materials. The study suggested that the added recyclates affect the overall properties, for example the tenacity decreases and the infrared crystallinity index shows a positive trend with increasing recyclate concentration. The researchers concluded that there is a tolerance level up to which the changes may be acceptable for many geotechnical applications. At least 20% w/w of the selected recyclate can be used in nonstabilised filaments with acceptable reduction in tensile properties, and with the addition of appropriate stabilisers the level of recyclate may be increased further.

Taking into account the durability aspects of the European Commission regulations, all buildings and construction materials' manufacturers have to comply with the requirements for CE marking, as CE marking for geotextiles has become mandatory from October 2002. Some of the important geotextile durability test methods are illustrated in Table 8. It must be noted that some of the discussions made in this Section and Section 2.3 have mainly focused on chemical durability of synthetic polymers that are generally used in the production of geotextiles. However, the mechanical durability of geotextiles such as creep, stress–strain behaviour, puncture etc. has been of significant importance and some of these properties have already been discussed (see Section 2.1).

4. Summary and perspectives

Geotextiles are widely used in various civil engineering and geotechnical applications and it is expected to be the fastest growing sector of the technical textiles market. In developed countries, polypropylene and polyester have been extensively used for the production of geotextiles primarily due to low cost and acceptable mechanical and chemical properties. However, natural fibre geotextiles are readily available in increasing amounts in Bangladesh, China, India and other developing countries, which have indigenous supplies of suitable natural fibres. In general, natural fibres are used for the short-term applications of geotextiles and subsequently degraded, whereas synthetic fibres have the desired properties fulfilling the long-term applications of geotextiles. In terms of fabric structure, nonwoven geotextiles have grown at a much faster rate than their woven counterparts, probably due to better compatibility with soil in terms of deformability. The selection of either woven or nonwoven fabrics is largely dictated by the properties of geotextiles fulfilling the requirements of desired applications through its functions. These properties should be tested in a manner which simulates the operational conditions in which geotextiles are expected to perform. For instance, the tensile properties should be determined under confined conditions of soil rather than testing through strip or grab methods that are generally carried out for other textile applications. Moreover, the commonly used geotextiles are highly porous in nature and therefore pore parameters and characteristics such as porosity, pore size distribution, pore connectivity etc. need to be determined by test methods that are universally acceptable.

The standardisation of properties and test methods can deliver a reliable technical data and design methodology for these materials, which will further result in the designing of geotextile by "design by function route". In this manner, the geotextile will be able to carry out its prescribed functions effectively during the project's lifetime and the durability of geotextiles can be determined. Perhaps, the combination of material and soil characteristics and process–structure–property relationship of geotextiles are of paramount importance for its design and durability. This suggests that there should be a clear understanding between the textile producers and the construction industry and it is necessary to transfer the information to textile professionals to realise the peculiarities of the civil engineering market.

References

[1] M.J. Denton and P.N. Daniels, *Textile Terms and Definitions*, 11th ed., The Textile Institute, Manchester, UK, 2002.
[2] N.W.M. John, *Geotextiles*. Blackie and Sons, Glasgow, Scotland, 1987.
[3] M. Pritchard, *Vegetable fibre geotextiles*, PhD dissertation. University of Bolton, Bolton, UK, 1999.
[4] A.R. Horrocks and S.C. Anand (eds.), *Handbook of Technical Textiles*, Woodhead, Cambridge, UK, 2000.
[5] W.K. Beckham and W.H. Mills, Eng. News-Rec. 115 (1935) pp. 453–455.
[6] J.P. Giroud, Geotext. Geomembranes 1 (1984) pp. 5–40.
[7] G.S. Hwang, C.K. Lu, M.F. Lin, B.L. Hwu and W.H. Hsing, Text. Res. J. 69 (1999) pp. 565–569.
[8] N.D. Williams and J. Luna, Geotext. Geomembranes 5 (1987) pp. 45–61.
[9] J.E. Fluet, J. Ind. Text. 14 (1984) pp. 53–64.
[10] Y. Wang, J. Ind. Text. 30 (2001) pp. 289–302.
[11] T.K. Ghosh, Geotext. Geomembranes 16 (1998) pp. 293–302.
[12] S. Adanur and T. Liao, Text. Res. J. 68 (1998) pp. 155–162.
[13] T. Liao, S. Adanur and J.-Y. Drean, Text. Res. J. 67 (1997) pp. 753–760.
[14] Y.E.El. Mogahzy, Y. Gowayed and D. Elton, Text. Res. J. 64 (1994) pp. 744–755.
[15] S.C. Anand, Indian J. Fibre Text. Res. 33 (2008) pp. 339–344.
[16] K.R. Lekha, Geotext. Geomembranes 22 (2004) pp. 399–413.
[17] A. Rawal and R.D. Anandjiwala, Geotext. Geomembranes 25 (2007) pp. 61–65.
[18] T.B. Ahn, S.D. Cho and S.C. Yang, Geotext. Geomembranes 20 (2002) pp. 135–146.
[19] K.R. Datye and V.N. Gore, Geotext. Geomembranes 13 (1994) pp. 371–388.
[20] S.R. Kaniraj and G.V. Rao, Geotext. Geomembranes 13 (1994) pp. 389–402.
[21] S.L. Lee, G.P. Karunaratne, S.D. Ramaswamy, M.A. Aziz and N.C. Das Gupta, Geotext. Geomembranes 13 (1994) pp. 457–474.
[22] K.R. Lekha and V. Kavitha, Geotext. Geomembranes 24 (2006) pp. 38–51.
[23] G.V. Rao, J.P. Sampath Kumar and P.K. Banerjee, Geotext. Geomembranes 18 (2000) pp. 367–384.
[24] T. Sanyal and K. Chakraborty, Geotext. Geomembranes 13 (1994) pp. 127–132.
[25] K. Slater, J. Text. I. 94 (2003) pp. 99–105.
[26] Siew A. Tan, G.P. Karunaratne and N. Muhammad, Geotext. Geomembranes 12 (1993) pp. 363–376.
[27] S.R. Ranganathan, Geotext. Geomembranes 13 (1994) pp. 421–433.
[28] K. Balan and G. Venkatappa Rao, *Erosion control with natural geotextiles*, in *Environmental Geotechnology with Geosynthetics*, G.V. Rao and P.K. Banerjee, eds., The Asian Society for Environmental Geotechnology and CBIP, New Delhi, 1996, pp. 317–325.
[29] M.J. Wall, G.C. Frank and J.R. Stevens, Text. Res. J. 41 (1971) pp. 38–43.
[30] R.B. Barnett and K. Slater, J. Text. I. 82 (1991) pp. 417–425.
[31] H. Ludewig, *Polyester Fibers*, Wiley, New York, NY, 1971.
[32] B.P. Corbman, *Textile: Fiber to Fabric*, McGraw-Hill, New York, 1975.
[33] T. Shah and S.C. Anand, Technical Textiles Markets, 2nd quarter (2002) pp. 55–56.
[34] A. Rawal, D. Moyo, V. Soukupova and R.D. Anandjiwala, J. Ind. Text. 36 (2007) pp. 207–220.
[35] S.J. Russell (ed.), *Handbook of Nonwovens*, Woodhead Publishing, Boca Raton, FL, 2007.
[36] P. Potluri, A. Rawal, M. Rivaldi and I. Porat, Composites: Part A 34 (2003) pp. 481–492.
[37] A. Rawal, *Generation of an expert system for the optimisation of net extrusion processes*, PhD dissertation. University of Bolton, Bolton, UK, 2002.

[38] David Rigby Associates. (2009). Retrieved January 20, 2009, from www.davidrigby-associates.com.

[39] K.B. Gautier, C.W. Kocher and J.-Y. Drean, Text. Res. J. 77 (2007) pp. 20–28.

[40] I. Juran and B. Christopher, J. Geotech. Eng-ASCE 115 (1989) pp. 905–926.

[41] H.I. Ling, T.H.W. Jonathan and F. Tatsuoka, Geotext. Geomembranes 11 (1992) pp. 185–219.

[42] A. El- Fermaoui and E. Nowatzki, *Effect of confining pressure on performance of geotextiles in soils*, in *Proceedings of Second International Conference on Geotextiles*, Las Vegas, Nevada, 1982.

[43] B.D. Siel, J.T.H. Wu and N.N.S. Chou, *In-soil stress-strain behaviour of geotextile*, in *Proceedings of Geosynthetics '87 Conference*, New Orleans, LA, 1987.

[44] J.T.H. Wu and C.K. Su, *Soil geotextile interaction mechanism in pullout test*, in *Proceedings of Geosynthetics '87 Conference*, New Orleans, LA, 1987.

[45] E. Guler and M.S.T. Biro, Geotext. Geomembranes 17 (1999) pp. 67–79.

[46] K.Z. Andrawes, A. McGown and M.H. Kabir, Geotext. Geomembranes 1 (1984) pp. 41–56.

[47] E. Leflaive, J.L. Paute and N. Segouin, *La mesure des caracteristiques detraction en vue des applications pratiques*, in *Proceedings of Second International Conference on Geotextiles*, Las Vegas, Nevada, 1982.

[48] J.P. Giroud, Geotext. Geomembranes 22 (2004) pp. 297–305.

[49] D. Cazzuffi and S. Venesia, *The mechanical properties of geotextiles: Italian standard and inter laboratory test comparison*, in *Proceedings of 3rd International Conference on Geotextiles*, IFAI, Vienna, Austria, 1986, pp. 695–700.

[50] A. McGown, K.Z. Andrawes, S. Pradhan and A.J. Khan, *Limit state design of geosynthetics reinforced soil structure*, in *Proceedings of 6th International Conference on Geosynthetics*, Atlanta, 1998.

[51] A. McGown and A.J. Khan, *The design of geosynthetic reinforced soil structures "Problems and solution"*, in *Proceedings of Pre-Conference Symposium on Ground Improvement and Geosynthetics*, Bangkok, Thailand, 1999.

[52] D.T. Bergado, S. Youwai, C.N. Hai and P. Voottipruex, Geotext. Geomembranes 19 (2001) pp. 299–328.

[53] J. Maneecharoen, *Factors affecting the laboratory testing of geotextiles*, Master's of Engineering dissertation (GE-96-13). School of Civil Engineering, Asian Institute of Technology, Bangkok, Thailand, 1997.

[54] S. Resl and G. Werner, *The influence of nonwoven needle-punched geotextiles on the ultimate bearing capacity of the subgrade*, in *Proceedings of the Third International Conference on Geotextiles*, Vienna, Austria, 1986.

[55] V.P. Murphy and R.M. Koerner, Geotech. Test. J. 3 (1988) pp. 167–172.

[56] R.F. Wilson-Fahmy, D. Narejo and R.M. Koerner, Geosynth. Int. 3 (1996) pp. 605–627.

[57] D. Narejo, R.M. Koerner and R.F. Wilson-Fahmy, Geosynth. Int. 3 (1996) pp. 629–653.

[58] A. Rawal, S.C. Anand and T. Shah, J. Ind. Text. 37 (2008) pp. 341–356.

[59] J.P. Giroud, *Designing with Geotextiles. Geotextiles and Geomembranes, Definitions, Properties and Design*, IFAI Publishers, Roseville, MN, 1984.

[60] F. Lhote and J.M. Rigo, *Study of the effect of the soil bearing capacity on the geotextiles puncture resistance*, in *Proceedings of International Nonwoven Fabrics Conference*, Atlanta, 1988.

[61] R. Antoine and L. Couard, Geotext. Geomembranes 14 (1996) pp. 585–600.

[62] K. Jha and J.N. Mandal, *A review of research and literature on the use of geosynthetics in the modern geotechnical world*, in *Proceedings of the First Indian Geotextiles Conference on Reinforced Soil and Geotextiles*, Bombay, India, 1988.

[63] M.H. Vidal, *The development and future of reinforced earth*, in *Proceedings of a Symposium on Earth Reinforcement at the ASCE Annual Convention*, Pittsburgh, PA, 1978.

[64] R.M. Koerner, *Designing with Geosynthetics*, Prentice Hall, Englewood Cliffs, NJ, 1986.

[65] H. Kabeya, A.K. Karmokor and Y. Kamata, Text. Res. J. 63 (1993) pp. 604–610.

[66] A.B. Fourie and K.J. Fabian, Geotext. Geomembranes 6 (1993) p. 275.

[67] A. Collios, P. Delmas, J.-P. Gourc and J.-P. Giroud, *The Use of Geotextiles for Soil Improvement*, ASCE National Convention, Portland, Oregon, 1980.

[68] E. Dembicki and P. Jermolowicz, Geotext. Geomembranes 10 (1991) pp. 249–268.

[69] I. Yogarajah and K.C. Yeo, Geotext. Geomembranes 13 (1994) pp. 43–54.

[70] A.K. Karmokar, H. Kabeya and Y. Tanaka, J. Text. I. 87 (1996) pp. 586–594.

[71] K. Farrag, Y.B. Acar and I. Juran, Geotext. Geomembranes 12 (1993) p. 133.
[72] M. Kharchafi and M. Dysli, Geotext. Geomembranes 12 (1993) pp. 307–325.
[73] M.J. Lopes and M.L. Lopes, Geosynth. Int. 6 (1999) pp. 261–282.
[74] R.M. Bakeer, A.H. Abdel-Rahman and P.J. Napolitano, Geotext. Geomembranes 16 (1998) pp. 73–85.
[75] S.M. Haeri, R. Noorzad and A.M. Oskoorouchi, Geotext. Geomembranes 18 (2000) pp. 385–402.
[76] H. Zhai, S.B. Mallick, D. Elton and S. Adanur, Text. Res. J. 66 (1996) pp. 269–276.
[77] A. Bouazza and S. Djafer-Khodja, Geotext. Geomembranes 13 (1994) pp. 807–812.
[78] K. Garbulewski, *Direct shear and pull-out frictional resistance at the geotextile and mud surface*, in *Proceedings of 4th International Conference Geotextile and Geomembranes & Related Products*, La Hague, France, 1990.
[79] K.D. Eigenbrod, J.P. Burak and J.G. Locker, Can. Geotech. J. 27 (1990) pp. 520–526.
[80] J.P. Giroud, J. Darrase and R.C. Bachus, Geotext. Geomembranes 12 (1993) pp. 275–286.
[81] G.A. Athanasopoulos, Geotext. Geomembranes 14 (1996) pp. 619–644.
[82] K.J. Fabian and A.B. Fourie, *Clay-geotextile interaction in soil tensile tests*, in *Proceedings of 1st Symposium on Earth Reinforcement Practice*, Fukuoka, Japan, 1988.
[83] S.A. Tan, S.H. Chew and W.K. Kong, Geotext. Geomembranes 16 (1998) pp. 161–174.
[84] Y. Wasti and Z.B. Ozduzgun, Geotext. Geomembranes 19 (2001) pp. 45–57.
[85] C.M. van Wyk, J. Text. I. 37 (1946) pp. T285–T292.
[86] A.E. Stearn, J. Text. I. 62 (1971) pp. T353–T360.
[87] G.A. Carnaby and N. Pan, Text. Res. J. 59 (1989) pp. 275–284.
[88] V.K. Kothari and A. Das, Indian J. Fibre Text. Res. 21 (1996) pp. 235–243.
[89] D.H. Lee and J.K. Lee, *Initial compressional behavior of fibre assembly*, in *Objective Measurement: Applications to Product Design and Process Control*, The Textile Machinery Society of Japan, Osaka, Japan, 1985, pp. 613–622.
[90] A. Rawal, J. Text. I. 100 (2009) pp. 28–34.
[91] V.K. Kothari and A. Das, Geotext. Geomembranes 11 (1992) pp. 235–253.
[92] V.K. Kothari and A. Das, J. Text. I. 84 (1993) pp. 16–30.
[93] V.K. Kothari and A. Das, Geotext. Geomembranes 13 (1994) pp. 55–64.
[94] J.P. Giroud, Mater. Construc. 14 (1981) pp. 257–272.
[95] S. Sengupta, P. Ray and P.K. Majumdar, Indian J. Fibre Text. Res. 30 (2005) pp. 389–395.
[96] A. Rawal, J. Text. I. 99 (2008) pp. 9–15.
[97] G.D. Hoedt, Geotext. Geomembranes 4 (1986) pp. 83–92.
[98] C.S. Wu, Y.S. Hong and R.H. Wang, Geotext. Geomembranes 26 (2008) pp. 250–262.
[99] S.Z. Cui and S.Y. Wang, Text. Res. J. 66 (1999) pp. 931–934.
[100] A. Sawicki and K.K. Frankowska, Geotext. Geomembranes 16 (1998) pp. 365–382.
[101] J.G. Zornberg, M. Asce, B.R. Blyer and J.W. Knudsen, J. Geotech. Geoenviron. 130 (2004) pp. 1158–1167.
[102] I. Fatt, Petroleum Transactions AIME 207 (1956) pp. 144–159.
[103] T. Komori and F. Makashima, Text. Res. J. 49 (1979) pp. 550–555.
[104] L. Rebenfield and B. Miller, J. Text. I. 86 (1995) pp. 241–251.
[105] A. Rawal, J. Text. I. 101 (2010) pp. 350–359.
[106] C.T.J. Dodson and W.W. Sampson, Appl. Math. Lett. 10 (1997) pp. 87–89.
[107] W.W. Sampson, J. Mater. Sci. 38 (2003) pp. 1617–1622.
[108] J. Castro and M.O. Starzewski, Appl. Math. Model. 24 (2000) pp. 523–534.
[109] S.K. Bhatia and J.L. Smith, Geotext. Geomembranes 13 (1994) pp. 679–702.
[110] B.S. Gerry and G.P. Raymond, Geotech. Test. J. 6 (1983) pp. 53–63.
[111] F. Saathoff and S. Kohlhase, *Research at the Franzius-Institute on geotextile filters in hydraulic engineering*, in *Proceedings of the Fifth Congress Asian and Pacific Region Division*, ADP/IAHR, Seoul, Korea, 1986.
[112] Y.H. Faure, J.P. Gourc and P. Gendrin, *Structural study of porometry and filtration opening size of geotextiles*, in *Geosynthetics: Microstructure and Performance, ASTM STP 1076*, I.D. Peggs, ed., ASTM International, Philadelphia, 1990, pp. 102–119.
[113] S.K. Bhatia, J.L. Smith and B.R. Christopher, Geosynth. Int. 3 (1996) pp. 85–105.
[114] S.K. Bhatia, J.L. Smith and B.R. Christopher, Geosynth. Int. 3 (1996) pp. 155–180.
[115] S.K. Bhatia, J.L. Smith and B.R. Christopher, Geosynth. Int. 3 (1996) pp. 301–328.
[116] D.J. Elton and D.W. Hayes, Geosynth. Int. 15 (2008) pp. 22–30.

[117] W. Dierickx, Geotext. Geomembranes 17 (1999) pp. 231–245.

[118] J. Mlynarek, Geotext. Geomembranes 2 (1985) pp. 65–77.

[119] A.H. Aydilek, S.H. Oguz and T.B. Edil, J. Comput. Civil Eng. 16 (2002) pp. 280–290.

[120] M.G. Gardoni and E.M. Palmeira, Geotechnique 52 (2002) pp. 405–418.

[121] E.M. Palmeira and M.G. Gardoni, Geotext. Geomembranes 20 (2002) pp. 97–115.

[122] A.B. Fourie and P.C. Addis, Geotext. Geomembranes 17 (1999) pp. 331–340.

[123] G.R. Fischer, R.D. Holtz and B.R. Christopher, *A critical review of geotextile pore size measurement methods*, in *Proceedings of the First International Conference "Geofilters,"* A.A. Balkema, Karlsruhe, Germany, 1992, pp. 83–89.

[124] A. Bouazza, Geotext. Geomembranes 22 (2004) pp. 531–541.

[125] H.I. Ling and F. Tatsuoka, Geotext. Geomembranes 12 (1993) pp. 509–542.

[126] J.P. Giroud, A. Zhao and G.N. Richardson, Geosynth. Int. 7 (2000) pp. 433–452.

[127] R.M. Koerner, J.A. Bove and J.P. Martin, Geotext. Geomembranes 1 (1984) pp. 57–73.

[128] E.M. Palmeira and M.G. Gardoni, Geosynth. Int. 7 (2000) pp. 403–331.

[129] G.S. Hwang, B.L. Hwu, W.H. Hsing and C.K. Lu, Geotext. Geomembranes 16 (1998) pp. 355–363.

[130] K.Y. Wei, T.L. Vigo, B.C. Goswami and K.E. Duckett, Text. Res. J. 55 (1985) pp. 620–626.

[131] S.A. Ariadurai and P. Potluri, Text. Res. J. 69 (1999) pp. 345–351.

[132] A. Rawal, J. Text. I. 97 (2007) pp. 527–532.

[133] N. Mao and S.J. Russell, J. Text. I. 91 (2000) pp. 235–243.

[134] N. Mao and S.J. Russell, J. Text. I. 91 (2000) pp. 243–258.

[135] N. Mao and S.J. Russell, J. Appl. Phys. 94 (2003) pp. 4135–4138.

[136] B.S. Gerry and G.P. Raymond, Geotech. Test. J. 6 (1983) pp. 181–189.

[137] S.M. Montgomery, K.L. Adams and L. Rebenfeld, Geotext. Geomembranes 7 (1988) pp. 275–292.

[138] S.M. Montgomery, B. Miller and L. Rebenfeld, Text. Res. J. 62 (1992) pp. 151–161.

[139] K.L. Adams and L. Rebenfeld, Text. Res. J. 57 (1987) pp. 647–654.

[140] R.A. Greenkorn, *Flow Phenomena in Porous Media*, Marcel Dekker, New York, 1983.

[141] D. Adolphe, M. Lopes and D. Sigli, Text. Res. J. 64 (1994) pp. 176–182.

[142] M.A. Saidi, J.Y. Drean and D. Adolphe, Text. Res. J. 69 (1999) pp. 10–15.

[143] A.R. Horrocks, *The Durability of Geotextiles*, EUROTEX 24 (1992).

[144] G.R. Koerner, G.Y. Hsuan and R.M. Koerner, J. Geotech. Geoenviron. 124 (1998) pp. 1159–1166.

[145] T.S. Ingold, *The Geotextiles and Geomembranes Manual*, Vol. 1, Elsevier, Oxford, MS, 1994.

[146] R.M. Koerner, *Durability and Aging of Geosynthetics*, Vol. 3, Elsevier, New York, 1989.

[147] R.M. Koerner, A.E. Lord and Y.H. Halse, Geotext. Geomembranes 7 (1988) pp. 147–158.

[148] H.Y. Jeon, S.H. Cho, M.S. Mun, Y.M. Park and J.W. Jang, Polym. Test. 24 (2005) pp. 339–345.

[149] A. Mathur, A.N. Netravali and T.D. O'Rourke, Geotext. Geomembranes 13 (1994) pp. 591–626.

[150] E. Richaud, F. Farcas, L. Divet and J.P. Benneton, Geotext. Geomembranes 26 (2008) pp. 71–81.

[151] E.M. Palmeira, A.F.M. Remigio, M.L.G. Ramos and R.S. Bernardes, Geotext. Geomembranes 26 (2008) pp. 205–219.

[152] Y. Halse, R.M. Koerner and A.E. Lord, Geotext. Geomembranes 6 (1987) pp. 295–305.

[153] S.R. Malkan and L.C. Wadsworth, International Nonwovens Bulletin 3 (1992) pp. 4–14.

[154] S.R. Malkan and L.C. Wadsworth, International Nonwovens Bulletin 4 (1992) pp. 24–33.

[155] S.R. Malkan, L.C. Wadsworth and C.R. Davey, International Nonwovens Journal 6 (1994) pp. 42–50.

[156] S.R. Malkan, L.C. Wadsworth and C.R. Davey, International Nonwovens Journal 6 (1994) pp. 50–57.

[157] S.R. Malkan, L.C. Wadsworth and C.R. Davey, International Nonwovens Journal 6 (1994) pp. 57–61.

[158] S.R. Malkan, L.C. Wadsworth and C.R. Davey, International Nonwovens Journal 6 (1994) pp. 61–70.

[159] J.W.S. Hearle, M.A.I. Sultan and S. Govender, J. Text. I. 67 (1976) pp. 373–376.

[160] J.W.S. Hearle, M.A.I. Sultan and S. Govender, J. Text. I. 67 (1976) pp. 377–381.

[161] J.W.S. Hearle, M.A.I. Sultan and S. Govender, J. Text. I. 67 (1976) pp. 382–386.

[162] R. Salvado, J. Silvy and J.Y. Drean, Text. Res. J. 76 (2006) pp. 805–812.

[163] J.W.S. Hearle and M.A.I. Sultan, J. Text. I. 58 (1967) pp. 251–265.

[164] J.W.S. Hearle, M.A.I. Sultan and T.N. Choudhari, J. Text. I. 59 (1968) pp. 103–116.

[165] A.T. Purdy, *Needle-punching*, *Textile Progress*, The Textile Institute, Manchester, UK, 1980.

[166] M. Miao, Text. Res. J. 74 (2004) pp. 394–398.

[167] A.K. Rakshit, A.N. Desai and N. Balasubramananian, Indian J. Fiber Text. Res. 15 (1990) pp. 41–48.

[168] A. Rawal and R.D. Anandjiwala, J. Ind. Text. 35 (2006) pp. 271–285.

[169] A. Rawal, A. Majumdar, S. Anand and T. Shah, J. Appl. Polym. Sci. 112 (2009) pp. 3575–3581.

[170] E. Leflaive, Geotext. Geomembranes 2 (1985) pp. 23–30.

[171] R. Koerner, *Should I Specify a Woven or Nonwoven?* Geotechnical Fabrics Report, IFAS, Minnesota, USA, 1984.

[172] S.M. Luettich, J.P. Giroud and R.C. Bachus, Geotext. Geomembranes 11 (1992) pp. 355–370.

[173] B.R. Christopher and G.R. Fischer, Geotext. Geomembranes 11 (1992) pp. 337–353.

[174] A.L. Rollin and G. Lombard, Geotext. Geomembranes 7 (1988) pp. 119–145.

[175] P.D.J. Watson and N.W.M. John, Geotext. Geomembranes 17 (1999) pp. 265–280.

[176] T.S. Ingold, Geotext. Geomembranes 2 (1985) pp. 31–45.

[177] P. Bertacchi and D. Cazzuffi, *The suitability of geotextiles as filters*, in *Proceedings of International Conference on Materials for Dams*, Monte Carlo, Monaco, 1984.

[178] J.P. Giroud, *Filter criteria for geotextile*, in *Proceedings of the Second International Conference on Geotextiles*, Las Vegas, Nevada, 1982.

[179] A.M. Elsharief and C.W. Lovell, Geotext. Geomembranes 14 (1996) pp. 601–617.

[180] L.F. Liu and C.Y. Chu, Geotext. Geomembranes 24 (2006) pp. 325–330.

[181] K.V. Harten, Geotext. Geomembranes 3 (1986) pp. 53–76.

[182] R.H. Chen, C.C. Ho and C.Y. Hsu, Geosynth. Int. 15 (2008) pp. 95–106.

[183] W. Schober and H. Teindl, *Filter criteria for geotextiles*, in *Proceedings of the Seventh European Conference on Soil Mechanics and Foundation Engineering*, Brighton, UK, 1979.

[184] J.A. Wates, *Filtration, an application of a statistical approach to filters and filter fabrics*, in *Proceedings of the Seventh Regional Conference for Africa on Soil Mechanics and Foundation Engineering*, The Netherlands, 1980.

[185] R.G. Carroll, Transportation Res. Rec. 916 (1983) pp. 46–53.

[186] Y.H. Faure, A. Baudoin, P. Pierson and O. Ple, Geotext. Geomembranes 24 (2006) pp. 11–20.

[187] L.J. Sansone and R.M. Koerner, Geotext. Geomembranes 11 (1992) pp. 371–393.

[188] Y.H. Faure, B. Farkouh, Ph. Delmas and A. Nancey, Geotext. Geomembranes 17 (1999) pp. 353–370.

[189] M. Sato, T. Yoshida and M. Futaki, Geotext. Geomembranes 4 (1986) pp. 223–240.

[190] D.B. Narejo, Geotext. Geomembranes 21 (2003) pp. 257–264.

[191] P.J. Black and R.D. Holtz, J. Geotech. Geoenviron. 125 (1999) pp. 404–412.

[192] R.D. Espinoza, Geotext. Geomembranes 13 (1994) pp. 281–293.

[193] J.P. Giroud and L. Noirey, Geotechnical Division ASCE, 107 (1981) pp. 1233–1254.

[194] R.D. Holtz and N. Sivakugan, Geotext. Geomembranes 5 (1987) pp. 191–199.

[195] R.V.V. Zanten (ed.), *Geotextiles and Geomembranes in Civil Engineering*, A.A. Balkema Publishers (Accord MA 02018), Rotterdam, Netherlands, 1986.

[196] R.D. Espinoza and P.J. Sabatini, Geosynth. Int. 15 (2008) pp. 350–357.

[197] B. Broms, Geotext. Geomembranes 5 (1987) pp. 17–28.

[198] R.A. Jewell, Geotext. Geomembranes 2 (1985) pp. 83–109.

[199] J.M. Rigo, Y. Mathieu, K. Smolders and E. Alexandre, *Geotextile Testing Inventory of Current Test Methods and Standards*, I.G.S. publication, Liege, Belgium, 1991.

[200] J.M. Rigo and Ch. Legrand, Geotext. Geomembranes 12 (1993) pp. 461–469.

[201] Y. Matichard, J.M. Rigo and Ch. Legrand, Geotext. Geomembranes 12 (1993) pp. 451–460.

[202] G. Mannsbart and S. Resl, Geotext. Geomembranes 12 (1993) pp. 441–450.

[203] T.S. Ingold, *Civil engineering requirements for long-term behaviour of geotextiles.* Seminar organized by RILEM, the International College of Building Science and International Geotextile Society, Saint-Remy-le's-Chevreuse, France, 4–6 November, 1986.

[204] N.A. Lebedev, Fibre Chem. 25 (1994) pp. 505–508.

[205] E.W. Brand and P.L.R. Pang, J. Geotech. Eng. 117 (1990) pp. 979–1000.

[206] A. McGown, K.A. Andrews and H. Al-Mudhaf, *Assessment of the effects of long-term exposure on the strength of geotextiles and geogrids*, in *Proceedings of Geosynthetics '95 Conference*, IFAI, Nashville, TN, 1995.

[207] L. Cassady and D.G. Bright, *Durability of geosynthetics exposed to nine years of natural weathering*, in *Proceedings of Geosynthetics '95 Conference*, IFAI, Nashville, TN, 1995.

[208] C.W. Hsieh and C.K. Lin, Transportation Res. Rec. 1849 (2003) pp. 221–230.

[209] J.R. Horlker, B. Vevers and J.O. Warwicker, Trans. Inst. Marine Eng. 97 (1984) pp. 161–167.

[210] C.C. Hsieh, J.B. Wang and Y.F. Chiu, Geosynth. Int. 13 (2006) pp. 210–217.

[211] J.D. Keenan, R.L. Steiner and A.A. Fungaroli, Journal of Waste Pollution Control Federation 56 (1984) pp. 27–33.

[212] C. Oman and P.A. Hynning, Environ. Pollut. 80 (1993) pp. 265–271.

[213] J.P. Gourc and Y. Faure, *Soil particles, water and fibres—a fruitful interaction now controlled*, in *Proceedings of the 4th International Conference on Geotextiles, Geomembranes and Related Products*, The Hague, The Netherlands, 1990.

[214] R.M. Koerner (ed.), *Biological activity and potential remediation involving geotextile landfill leachate filters*, in *ASTM STP 1081*, American Society for Testing and Materials (ASTM) (Special Technical Publications), Philadelphia, PA, 1990.

[215] J. Lafleur, Geotext. Geomembranes 17 (1999) pp. 299–312.

[216] R. McIsaac and R.K. Rowe, Can. Geotech. J. 42 (2005) pp. 1173–1188.

[217] R. McIsaac and R.K. Rowe, ASCE J. Geotech. Geoenviron. 133 (2007) pp. 1026–1039.

[218] A.R.L. Silva, E.M. Palmeira and G.R. Vieira, *Large filtration tests on drainage systems using leachate*, in *Proceedings of the 4th International Congress on Environmental Geotechnics*, Rio de Janeiro, Brazil, 2002.

[219] Geofabrics. *The durability of GEO-fabrics non-woven geotextiles* (Tech. Rep. No. 3.14). International Geosynthetics Society, Leeds, UK, 2005.

[220] M. Akkapeddi, B.V. Biskirk, B. Mason, C.D. Chung and X. SwamiKannu, Polym. Eng. Sci. 35 (1995) pp. 72–78.

[221] R. Pfaender, H. Herbst, K. Hoffmann, B. Klingert, F. Sitek and T. Cooper, Kunststaffe/Plast Eur. 85 (1995) pp. 81–84.

[222] K. Rebeiz, D.W. Fowler and D.R. Paul, Trends Polym. Sci. 1 (1993) pp. 315–321.

[223] P.J. Davies and A.R. Horrocks, Text. Res. J. 70 (2000) pp. 363–372.

Taylor & Francis
Author Services

publish with us

The **Taylor & Francis Author Services** department aims to enhance your publishing experience as a journal author and optimize the impact of your article in the global research community. Assistance and support is available – from preparing the submission of your article through to setting up citation alerts post-publication on Informaworld.

Our **Author Services** department can provide advice on how to:

- **Direct your submission to the correct journal**
- **Prepare your manuscript according to the journal's requirements**
- **Maximize your article's citations**
- **Submit supplementary data for online publication**
- **Submit your article online via ScholarOne Manuscripts**
- **Apply for permission to reproduce images**
- **Prepare your illustrations for print**
- **Track the status of your manuscript through the production process**
- **Return your corrections online**
- **Purchase reprints through *Rightslink***
- **Register for article citation alerts**
- **Take advantage of our i*OpenAccess* option**
- **Access your article online**
- **Benefit from rapid online publication via i*First***

For further information go to:
www.tandf.co.uk/journals/authorservices

Or contact:

Author Services Manager, Taylor & Francis Group,
4 Park Square, Milton Park, Abingdon, Oxon OX14 4RN UK
Email: **authorqueries@tandf.co.uk**

Taylor & Francis
Taylor & Francis Group

Routledge
Taylor & Francis Group

Psychology Press
Taylor & Francis Group

Powered by
informaworld

Taylor & Francis
Taylor & Francis Group

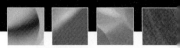

International Journal of Fashion Design, Technology and Education

New journal in 2008

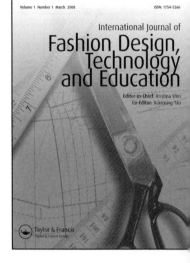

EDITOR:

Kristina Shin, *The Hong Kong Polytechnic University, China*

International Journal of Fashion Design, Technology and Education aims to provide a high quality peer-reviewed forum for research in fashion design, pattern cutting, apparel production, manufacturing technology and fashion education. The Journal will encourage interdisciplinary research and the development of an academic community which will share newly developed technology, theory and techniques in the fashion and textile industries, as well as promote the development of excellent education practice in the clothing and textile fields.

Contributions suitable for this new journal should fall into one of the following three categories:

- Research papers presenting important new findings
- Technical papers describing new developments or innovation
- Academic discussion papers dealing with medium to long-term trends and predictions.

To receive the table of contents for *International Journal of Fashion Design, Technology and Education* visit the journal homepage at www.tandf.co.uk/journals/tfdt

Submit your papers via Manuscript Central at http://mc.manuscriptcentral.com/tfdt

---

To sign up for tables of contents, new publications and citation alerting services visit **www.informaworld.com/alerting**

*e*updates
Taylor & Francis Group

Register your email address at **www.tandf.co.uk/journals/eupdates.asp** to receive information on books, journals and other news within your areas of interest.

Powered by
informaworld

For further information, please contact Customer Services at either of the following:

T&F Informa UK Ltd, Sheepen Place, Colchester, Essex, CO3 3LP, UK
Tel: +44 (0) 20 7017 5544 Fax: 44 (0) 20 7017 5198
Email: subscriptions@tandf.co.uk

Taylor & Francis Inc, 325 Chestnut Street, Philadelphia, PA 19106, USA
Tel: +1 800 354 1420 (toll-free calls from within the US)
or +1 215 625 8900 (calls from overseas) Fax: +1 215 625 2940
Email: customerservice@taylorandfrancis.com

View an online sample issue at:
www.tandf.co.uk/journals/tfdt